S0-CBI-346

Fundamental Measures and Constants for Science and Technology

Author:
Frederick D. Rossini
Professor of Chemistry
Rice University
Houston, Texas

QC
39
R66
1974

Published by

CRC PRESS, Inc.
18901 Cranwood Parkway · Cleveland, Ohio 44128

310898

Library of Congress Cataloging in Publication Data

Rossini, Frederick Dominic, 1899-
Fundamental measures and constants for science and technology.

Includes bibliographies.
1. Units. 2. Physical measurements. I. Title.
QC39.R66 389 74-14759
ISBN 0-87819-051-1

This book represents information obtained from authentic and highly regarded sources. Reprinted material is quoted with permission, and sources are indicated. A wide variety of references are listed. Every reasonable effort has been made to give reliable data and information, but the author and the publisher cannot assume responsibility for the validity of all materials or for the consequences of their use.

All rights reserved. This book, or any parts thereof, may not be reproduced in any form without written consent from the publisher.

© 1974 by CRC Press, Inc.

International Standard Book Number 0-87819-051-1

Library of Congress Card Number 74-14759
Printed in the United States

Dedicated

to

ANNE L. ROSSINI

My wife and companion of many travels
on the highways of science

THE AUTHOR

Frederick D. Rossini is professor of chemistry at Rice University, Houston.

Dr. Rossini received his B.S. in Chemical Engineering and M.S. in Science from Carnegie Institute of Technology, and his Ph.D. in Physical Chemistry from the University of California, Berkeley. He is the recipient of seven honorary degrees.

Dr. Rossini is an authority in the fields of thermodynamics and thermochemistry, numerical data for science and technology, and physical chemistry of petroleum and hydrocarbons. Dr. Rossini joined the staff of the National Bureau of Standards in 1928, and from 1936 to 1950 served as chief of the Section on Thermochemistry and Hydrocarbons. He was Silliman Professor, head of the department of chemistry, and Chemical and Petroleum Research Laboratory director at Carnegie Institute of Technology from 1950 to 1960. From 1960 to 1967, Dr. Rossini was dean of the College of Science and associate dean of the Graduate School at the University of Notre Dame. He served as vice president for research at Notre Dame from 1967 to 1971, when he went to Rice University.

Dr. Rossini is the author of over 240 publications, including 10 books.

PREFACE

This book on the fundamental measures and constants of science is based on my experiences of nearly 50 years devoted to researches and writings related to accurate measurements of thermodynamic and physical properties of chemical substances, particularly hydrocarbons and related compounds, and to the problems of numerical data for science and technology. It has been my privilege to have firsthand contact with some of the world's experts on the fundamental units of measurement, the scale of temperature, the scale of pressure, the scale of atomic masses, and the fundamental physical constants. The need to present the results of experimental measurements in a form and manner that will maximize their usefulness to scientists and engineers has been for me a continuing challenge, requiring adequate knowledge of and familiarity with the foregoing topics. Such matters are normally touched upon only casually and peripherally in regular courses of study, so that one is expected to develop knowledge of them on his own.

In 1972 and 1973, I gave detailed lectures on the above topics at Rice University, and in 1973 as the Strosacker Visiting Professor of Science at Baldwin-Wallace College. Having done this, it appeared desirable that the material presented in these lectures be put down in written form for the benefit of the many persons interested whom I cannot reach in the classroom. Each topic is presented in a simple, clear, and straightforward manner, including a brief history along with the present status.

It is my hope that this collection of information on the fundamental measures and constants of science will be really useful to working scientists and engineers as well as to undergraduate and graduate students in science and engineering. In particular, I hope that this book will provide for them, in convenient form, information that will ensure the reliability and maximize the effectiveness of their work in the area of measurements.

For the latest up-to-date information on the several topics, beyond material available in the open literature, I am greatly indebted to my friends in the several areas, whose names are indicated at the appropriate places in the text. I would greatly appreciate being informed of any errors of commission or omission in this book.

Frederick D. Rossini
April, 1974.

TABLE OF CONTENTS

Chapter 1

INTRODUCTION

Science is based upon observation and measurement. One of the most important capabilities possessed by man is the ability to measure. The better and more accurately we can observe and measure, the better and more accurately we can describe the phenomena of nature, develop theories to explain the natural state of things, and guide ourselves to more fruitful observations and measurements. The advance of science is proportional to the extent to which we have quantitative knowledge of the dimensions of material things, of the rates at which phenomena occur, of the forces that hold entities together, and of the changes in energy accompanying natural or man-made processes. With more knowledge, we are able to devise theories to correlate hitherto unrelated observations. In time, as theories become established, mountains of observational data can be replaced by a few simple formulas.

In the early beginnings of science, simple words and simple measures were adequate. Then, as science developed, with need for higher precision and accuracy, more precise words and measures became necessary to record observations and communicate the results. Today, the communication of scientific observations has become a highly complex and very important operation. One of the big problems that we have in science today is to communicate the results of observations and measurements in a manner that will be fully understood by other scientists.

Not infrequently, the full value of the measurements arising from a given investigation is not recovered because the Principal Investigator has not been sufficiently aware of the relation between his measurements and similar ones of others, and of the connection of such measurements with related other quantities. The resulting report or publication may be written in a manner that leads to less than full understanding by others in the same discipline. Communication without understanding can lead to misinterpretation and unnecessary and costly repetition of measurements. It is the obligation of each investigator to place his observations and measurements on a solid foundation by appropriate linkage with the fundamental units of measurement and proper use of appropriate values of the fundamental units of measurement involved. The scientific and technical literature contains many examples of reports of investigations that are woefully inadequate in such matters, as well as others that are prime examples of excellence.

We have the problem of communicating among scientists of the same country and then among scientists of different countries. Whether the communication is among scientists of the same discipline in the same country, among scientists of different disciplines in the same country, among scientists of the same discipline in different countries, or among scientists of different disciplines in different countries, it is important that the words, terms, and symbols used have the same meaning at both ends of the chain of communication. This means appropriate coordination of such matters by a national body in a given country and by an international body for all countries.

Within the past 150 years the speed of communication in the world has increased about several million times, and the speed of travel about several thousand times. This has brought all peoples of the world, including scientists and engineers, much closer together. This proximity in terms of time of communication and travel brings with it the need to communicate across national boundaries with adequate understanding.

Each of us can do his part in the scientific-technological endeavor by observing carefully, measuring with proper instruments of appropriate precision and accuracy, and then communicating the results in terms, units, and symbols that have international acceptance in the given disciplines of science and engineering.

Another very important reason for the spread of knowledge of the fundamental measures and constants of science is that the advance of science and technology in the service of mankind is dependent upon continued improvement in the precision and accuracy of the measurement and control of the variables of length, mass, time, temperature, pressure, and combinations of them. Our manufacturing industries require increasingly more efficient and finer control of their processes in order to compete successfully in the world markets. Similarly, many industries are now controlling new variables, representing complex combinations of the fundamental measures, that permit the production of new products hitherto undreamed of.

In the last century, Lord Kelvin wrote the following:

> I often say that when you can measure what you are speaking about, and express it in numbers, you know something about it; but when you cannot measure it, when you cannot express it in numbers, your knowledge is of a meager and unsatisfactory kind; it may be the beginning of knowledge, but you have scarcely, in your thoughts, advanced to the stage of science, whatever the matter may be.

Chapter 2

THE FUNDAMENTAL UNITS OF MEASUREMENT

A. INTRODUCTORY COMMENTS

In this chapter, we discuss the units of length, mass, and time, giving a brief history of them, describing the present standards for them, showing how they are maintained at the international and national levels, and indicating how they are carried down to the working and living levels. These standards are related to the measuring instruments used in laboratories, industry, commerce, business, sports, recreation, and the home.

Our yardsticks, our meter sticks, our balances and weighing machines, and our watches and clocks are tied through a chain of connections, some short and some long, to the basic units of length, mass, and time as defined by international agreement. The fundamental units of measurement are really of concern to everyone, not only in science, engineering, industry, and business, but also in our everyday experiences as individuals, in the market place and in the home. Even the housewife has a big stake in the matter of weights and measures.

B. THE UNIT OF LENGTH

Early records on units of length in the various countries of the world are most interesting and replete with native customs.

The cubit (of somewhat varying sizes) appears to be the name most frequently used in ancient history for the unit of length. The cubit is related to the length of the arm from the tip of the middle finger to the elbow. In Egypt, the cubit (52.4 cm) was used beginning with the time of the predynastic royal tombs. Variations of this cubit were found in Babylon (53.1 cm), in Asia Minor (52.2 to 53.2 cm), in Jerusalem (52.2 cm), in early Britain (52.2 cm), and in early stone buildings in what is now New Mexico, U.S. (52.5 cm).

Another early unit was about 3/5 of a cubit, found in Athens (31.6 cm), in Aigina (31.5 cm), in Miletos (31.8 cm), in Olympia (32.1 cm), in Etruria (31.6 cm), and in medieval Britain (31.7 cm). Another unit was about 2/3 of a cubit, used in Pergamon (35.1 cm). The short cubit (about 6/7 of a regular cubit) was used also in Egypt (45.0 cm) and in Jerusalem (44.7 cm). The Greek cubit has a special value (46.3 cm).

The digit was taken as 1/40 of the diagonal of a square, one cubit (52.4 cm) on a side, making the digit 1.85 cm.

Taking 2/3 of the Greek cubit (46.3 cm) produced the Greek foot (30.9 cm). The Greek foot in greater use was found to be shorter (29.5 cm). Similar units were Italic (29.6 cm), Rome (29.5 to 30.0 cm), Etrusca (29.4 cm), Stonehenge (29.7 cm), and other stone circles and hill figures (29.5 cm).

Another widespread measure found has a different unit: in early Egypt (33.8 cm), in Asia Minor (33.9 cm), in Greece (33.9 cm), at Lachish (900 B.C.) (33.5 cm), in Syria (620 A.D.) (33.6 cm), in medieval Britain as the commonest building unit (33.5 cm), and in some early French architecture (33.1 cm).

Another unit was used in Persepolis (48.8 cm), in the tower of Babylon (49.5 cm), in Asia Minor (49.0 cm), in early Assyria (50.7 cm), in Khorsabad (54.9 cm), in Phrygia (55.4 cm), in Lucania, Italy (55.5 cm), in late Egypt (53.6 to 54.2 cm), and in Persia (54.4 cm). Another important unit was used in Phoenicia (56.4 cm) and in Carthage and Sardinia (56.3 to 56.7 cm).

Measures of volume in the ancient systems developed independently of the units of length. Examples are the Egyptian "hen" (477 cm^3), the Syrian "kotyle" (341 to 354 cm^3), the Syrian "log" (544 cm^3), the Phoenician "log" (508 cm^3), the Babylonian "log" (541 cm^3), the Jewish "log" (544 cm^3), the Attic "kotyle" in Egypt (285 cm^3), the Persian "kapetis" (1,221 cm^3), and the Roman "amphora" (25.7 to 29.9 x 10^3 cm^3).

In Britain, the inch was originally taken as the length covered by three barleycorns, round and dry, laid end to end; the fathom was taken as the length from tip to tip of the fingers, with hands and arms outstretched; the yard was one half of the fathom.

In Germany, there is a record in the 16th Century showing the establishment of the German "rute" or "rod" as the length covered by the feet of 16 men, standing together in a line, toe to heel, selected as the men issued from church on a Sunday morning.

The early units of measure used in the United States were, of course, inherited from the British system, which had been used throughout the 13 colonies in America.

Following are some early notes on units of length and capacity in Britain. In 1439, the "yard and handful," or the "40-inch ell," was abolished. The "yard of Henry VII" (35.963 in.) was abolished in 1527. In 1553, the "yard and inch," or the "37-inch ell," was abolished. A "cloth ell" (45 in.) was used until 1600. Early measures of capacity included the "Winchester bushel" of Henry VII, the "ale gallon" of Henry VII, and the old Queen Anne "wine gallon" of 1707 (231 in.^{3}), which became the U.S. gallon.

Some of the units of length used in Britain at various times include the following (the number following the name is the nominal equivalent in inches): mil, 0.001; point, 1/72; line, 1/12; barleycorn, 1/3; palm, 3; hand, 4; span, 9; cubit, 18; pace, 30. Similarly, we have for longer units the following (the number is the nominal equivalent in feet): fathom, 6; rod, 16.5; rope, 20; chain, 66; skein, 360; furlong, 660; cable, 720; mile, 5,280; knot (nautical mile), 6,080; league, 15,840 (3 mi). Most of these units were also used in the American colonies and later in the U.S.

In 1758–60, a new British standard yard was constructed by direction of the Houses of Parliament in London. By the Weights and Measures Act of Parliament in 1878, the imperial yard was defined as the distance at 62°F between the axes of two lines traced on gold plugs set in a bronze bar preserved at the Standards Department of the Board of Trade in London. The legal equivalent of this was specified then as 0.9143992 m. However, later measurements by the British National Physical Laboratory showed that the yard as defined above was actually 0.9143987 m. The foot and the inch were taken as 1/3 and 1/36, respectively, of the imperial yard.

In 1790 in the U.S. President Washington suggested to the Congress that the United States should set up its own system of weights and measures. A report by the then Secretary of State, Thomas Jefferson, recommending a basic unit of length from which units of area, volume, etc., could be derived was accepted by the Congress, but, in spite of prodding by President Washington, the report was never implemented.

Nothing significant was done until 1816 when President James Madison reminded Congress that it was important that a uniform system of weights and measures be established. The U.S. Senate responded in the following year by passing a resolution requesting the Secretary of State to reinvestigate the problem. Four years later, in 1821, came the "Report upon Weights and Measures" submitted by Secretary of State John Quincy Adams. Adams' report included the following message:

> Weights and Measures may be ranked among the necessaries of life to every individual of human society. They enter into the economical arrangements and daily concerns of every family. They are necessary to every occupation of human industry; to the distribution and security of every species of property; to every transaction of trade and commerce; to the labors of the husbandman; to the ingenuity of the artificer; to the studies of the philosopher; to the researches of the antiquarian; to the navigation of the mariner, and the marches of the soldier; to all exchanges of peace, and all the operations of war. The knowledge of them, as in established use, is among the first elements of education, and is often learned by those who learn nothing else, not even to read and write. This knowledge is riveted in the memory by the habitual application of it to the employment of men throughout life.

Adams' report gave the following possible lines of action: (1) to adopt, in all its essential parts, the then-new French (metric) system; (2) to restore and perfect the old English system; (3) to devise and establish a new combined system by adapting parts of each system; (4) to adhere, without any innovation whatever, to the existing system – merely fixing the standards.

Adams himself preferred a two-stage approach: (1) standardization and approval of the customary familiar English units followed by (2) negotiations with France, Britain, and Spain to establish a uniform international system of measurement.

Adams' recommendations were practical in the sense of having some chance of approval, in view of the fact that by 1821 most states had already enacted laws specifying the English units of measure, and a sudden contrary national law might involve the problem of State's Rights. Further, it was a fact that the preponderance of United States trade in 1821 was still with Britain and that the U.S. was bounded by Canada and then-Spanish possessions. Congress took no action on Adams' report.

In 1832, the U.S. Department of the Treasury adopted the English standards of length and mass to meet the needs of customs houses.

In 1863, President Lincoln formed the National Academy of Sciences to advise the government on all technical matters. The Secretary of the Treasury appointed a committee, chaired by the eminent physicist, Joseph Henry, to reconsider the matter of weights, measures, and coinage. The committee issued its report 2 years later, favoring adoption of the French metric system.

In 1866 the newly appointed Committee on Coinage, Weights, and Measures of the U.S. House of Representatives, under the chairmanship of Congressman John A. Kasson, reported favorably on three bills dealing with the metric system. These were eventually passed by the Congress. One of the bills specified the metric equivalents of the English units used in the United States and made legal, though not compulsory, the use of metric weights and measures. Another bill directed the Postmaster General to distribute metric postal scales to all post offices handling foreign mail. The third bill directed the Secretary of the Treasury to provide each state with one set of metric standards. The following is quoted from the first of these bills:

> It shall be lawful throughout the United States of America to employ the weights and measures of the metric system; and no contract or dealing, or pleading in any court, shall be deemed invalid or liable to objection because the weights and measures expressed or referred to therein are weights and measures of the metric system.

In 1875, after 5 years of meetings in Paris, 17 nations signed the Treaty of the Meter. This treaty and convention accomplished several objectives: (a) the description of the metric system was clarified and reformulated to make the standards of the metric system more accurate; (b) provision was made for the construction of new standards of measurement; (c) provision was made for the distribution of accurate copies of these standards to the participating countries; (d) the International Bureau of Weights and Measures was created to serve as a world repository and laboratory, located at Sevres, near Paris, on a piece of international territory donated by France; (e) provision was made for continuing international conferences and action on weights and measures.

The present arrangements for international collaboration on weights and measures has the following pattern. The International Bureau of Weights and Measures, which aims to ensure worldwide uniformity of measurements by maintaining the international standards and carrying on comparisons of national and international standards, is under the cognizance of the International Committee on Weights and Measures which develops recommendations to be placed before the International General Conference on Weights and Measures, the top body in the enterprise. The General Conference consists of delegates from each member country of the Convention of the Meter and meets at least once every 6 years. The International Committee on Weights and Measures consists of 18 members, each from a different country, and meets at least once every 2 years. This International Committee has "consultative committees" for Electricity, Photometry, Thermometry, Definition of the Meter, Definition of the Second, and Standards for Measuring Ionizing Radiations and Units.

In 1889 the prototype copies of the international standard meter bar and international standard kilogram were completed and the U.S. received its copies. In 1893 the U.S. Secretary of the Treasury issued an administrative order declaring these new metric standards to be the fundamental standards of length and mass for the U.S. This meant that the customary units of length (inch, foot, yard, etc.) and of mass (pound) were defined in terms of the metric units:

1 yd	=	3,600/3,937 m	=	0.91440183 m
1 ft	=	1/3 yd	=	30.480061 cm
1 in.	=	1/36 yd	=	2.5400508 cm

This placed the U.S. on the basis of the metric system, although no effort was made towards a practical conversion of the day-to-day activities of the people to the units of the metric system. (The problem of the practical conversion to the metric system of the government, commercial, industrial, engineering, and personal measurement activities in the U.S. is discussed in the following chapter.)

In 1959 the U.S. in concert with the United Kingdom, Canada, Australia, New Zealand, and South Africa, agreed on precisely uniform definitions of the yard and the pound in terms of the metric equivalents. In the U.S. this was accomplished by a joint communique of the National Bureau of Standards and the U.S. Coast and

Geodetic Survey; with the approval of the Secretary of Commerce. The equivalent of the unit of length was given as:

1 yd = 0.9144 (exactly) m, so that
1 in. = 2.54 (exactly) cm, and
1 ft = 30.48 (exactly) cm

This changed the 1893 equivalent by two parts per million.*

Previous to 1889 the international meter bar was an "end standard," made of a platinum rod. An "end standard" is one where the given length is determined by the distance between the two parallel plane ends of the rod. (It should be noted that the size of the meter was originally selected so as to be one ten-millionth part of the quadrant of the earth's meridian passing through the poles and intersecting the equator at the earth's surface.)

In 1890, when the U.S. received its prototypes of the international standards, the international meter bar had been changed to a "line standard," made of an alloy consisting of platinum with 10% iridium by weight. A "line standard" is one where the given length is determined by the distance between the centers of two fine lines, cut parallel to each other, transversely on the rod. This rod had a "Tresca," or modified X, cross section, for resistance against deflection, with overall dimensions of 2 X 2 cm, as shown in Figure 2.1.

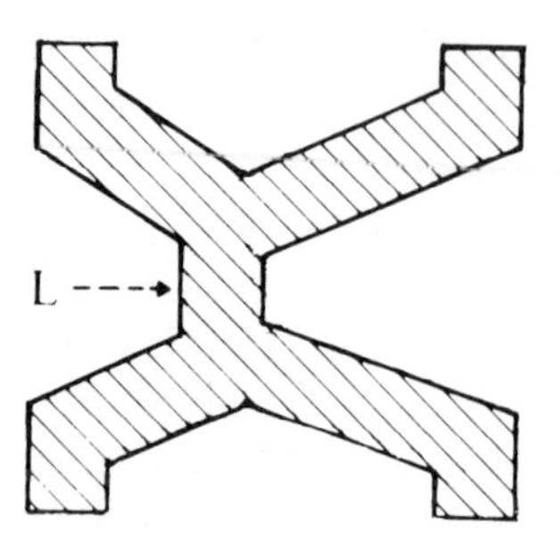

FIGURE 2.1. Cross section (Tresca) of the national standard meter bar of 1897. Two microscope lines were engraved on the measuring axis of the bar, one at each end, as indicated by L. (From Weights and Measures, in *Encyclopaedia Britannica,* Vol. 15, 14th ed., Encyclopaedia Britannica, New York, 1929, 135. With permission.)

In 1927 the Seventh International General Conference on Weights and Measures made the specifications for the international standard meter bar much more definite, as follows:

> The unit of length is the meter, defined by the distance, at the temperature of melting ice, between the centers of two lines traced on the platinum-iridium bar deposited at the International Bureau of Weights and Measures, and declared prototype of the meter by the First General Conference on Weights and Measures, this bar being subjected to normal atmospheric pressure and supported by two rollers, at least one centimeter in diameter, situated symmetrically in the same horizontal plane and at a distance of 572 millimeters from each other.

In discharging its obligations as the custodian and monitor of the unit of length, the National Bureau of Standards has maintained in its vaults the following units of length (obtained from appropriate government agencies which had received them before the establishment of the National Bureau of Standards): "Arago Platinum Meter," purchased from France in 1821; "Low Moor Iron Yard No. 57," copy of the British Imperial Yard, obtained as a gift from Britain in 1856; "Bronze Yard No. 11," copy of the British Imperial Yard, which was obtained as a gift from Britain in 1856 and served as the United States standard until 1893; "Committee Meter-Iron," copy of the first metric standard brought to the United States in 1905 and used by the U.S. Coast Survey from 1817 to 1890; "Prototype Meter No. 12," obtained from the International Bureau of Weights and Measures in 1890; "Prototype Meter No. 27," which was the reference standard of length for the U.S. from 1890 to 1960.

Beginning in the early part of this century, Michelson, in the U.S., had shown that it would be possible to base the international unit of length on the wavelength of selected monochromatic radiation. Work along these lines was also carried out by Fabry and Perot in France. The preferred radiation for this purpose then was principally the red line of cadmium. These investigators showed that, by comparison with the red line of cadmium, the international standard meter bar was unchanged, over a period of 15 years, to 0.1 ppm.

In 1927 the International General Conference adopted as an alternative and provisional defini-

*It should be noted that the geodetic survey records of the U.S. Coast and Geodetic Survey, maintained in terms of the equivalent of 1893, are exempted from this change until the time is propitious to readjust the basic geodetic survey networks to the new system.

tion of the meter the following: the meter is equivalent to 1,553,164.13 wavelengths of the red light emitted by a cadmium vapor lamp excited under certain specified conditions. It was taken that the uncertainty in this definition was 0.1 ppm.

But it was felt that much more experimentation was needed before the international standard meter bar could be abandoned completely. By about 1955, much of the needed experimental work was done, using principally the orange-red monochromatic radiation from the pure nuclide, Krypton-86

In October, 1969, the International General Conference on Weights and Measures made a new definition of the unit of length, in terms of the wavelength of monochromatic radiation from isotopically pure Kr-86, and gave up the international meter bar as the standard. They stated, "The unit of length is the meter, m, which is equal exactly to 1,650,763.73 wavelengths of light in vacuo produced by the unperturbed transition $2p_{10} - 5d_5$ in the pure nuclide, Krypton-86." (This is the orange-red radiation of Krypton.) Assuming an uncertainty of one half unit in the last figure written, this definition corresponds to an uncertainty of 0.003 ppm. This is to be compared with the corresponding uncertainty of 0.1 ppm attainable with the line-standard international meter bar.

The foregoing is the present international metric standard for the unit of length and is the legal basis of the system in the U.S. even before practical conversion to the metric system.

The new definition of the unit of length in terms of the wavelength of the monochromatic radiation from pure Krypton-86 eliminates world dependence on the security and validity of the international standard Pt-Ir bar maintained in the vaults of the International Bureau of Weights and Measures at Sevres. Also, the new definition eliminates the basic need for intercomparisons between national working standard meter bars, the national prototype bars, and, eventually, the international meter bar at Sevres.

At this point we can clear up a problem relating to the liter. The liter had originally been defined as the volume occupied by 1,000 g of water at its temperature of maximum density, near 4°C, at a pressure of 1 atm. When first defined, and confirmed by the Convention of the Meter in 1875, the liter was believed to be almost exactly 1,000 cm^3. Following an extensive investigation, the International Bureau of Weights and Measures reported, in 1910, the following:

1 liter = 1,000.027 cm^3.

In the International Critical Tables[1] in 1926, the relation was changed slightly to

1 liter = 1,000.028 cm^3.

This difference of 28 ppm between the milliliter and the cubic centimeter is small and can be neglected in many investigations. But there are many cases where this difference is very significant. Values of densities given to 1 to 10 ppm were required to be carefully and explicitly expressed as grams per milliliter or grams per cubic centimeter, depending upon the unit selected.

In 1964 the International General Conference on Weights and Measures eliminated this problem in the future by redefining the liter, independently of the properties of water, simply as the equivalent of one cubic decimeter:

1 liter = 1,000 (exactly) cm^3.

This means that values of density of high precision appearing in the literature prior to 1965, when expressed in grams per milliliter, require conversion to grams per cubic centimeter by dividing by 1.000028. Henceforth, values of density should normally be expressed in grams per cubic centimeter or a directly related quantity.

C. THE UNIT OF MASS

As in the case of units of length, units of mass in ancient times were many and varied and coupled with local customs. Examples are: in Palestine, the "shekel" or "peyem" (7.5 to 8.1 g), in Syria, the "manek" (408 g), in Persia, the "karasha" (834 g), in Egypt, the "gedet" (9.33 g), in Rome, the "libra" (327 g), and in Alexandria, the "sela" (14.3 g).

As in the case of units of length, the units of weight in the American colonies were those of Britain, where three different "English" systems were used:

Avoirdupois for general use
Apothecary for drugs
Troy for precious metals

In the British avoirdupois system, the various units used included the following (the number following the name gives the equivalent in pounds): dram, 1/256; ounce, 1/16; stone, customary, 8; stone, legal, 14; quarter, 28; cental, 100; hundred-weight, 112; ton, 2,240.

In the avoirdupois system in the U. S., similar units were used, as follows (with the number following the name being the equivalent in pounds): dram, 1/256; ounce, 1/16; hundred-weight, 112; ton, 2,000; long ton, 2,240. The troy and apothecary systems of the U.S. were the same as those of Britain. Following are the names and equivalents of these, in grains: for the troy system: pennyweight, 24; ounce, 480; pound, 5,760. For the apothecary system: scruple, 20; dram, 60; ounce, 480; pound, 5,760. In converting the foregoing, 1 lb avoirdupois is taken equivalent to 7,000 grains.

As previously reported in the preceding section on the discussions of the unit of length, the U.S. signed the Treaty or Convention of the Meter in 1875, along with 16 other countries. This Convention covered the unit of mass as well as the unit of length, and was an important step following the legalization (but not compulsory use) of the metric system by the U.S. Congress in 1866 and the accompanying definition of the yard and pound in terms of metric equivalents. It was in 1890 that the U.S. received its prototype of the international kilogram along with the international meter.

Earlier, the legal basis in the U.S. had been a prototype of the British imperial pound, which was a cylinder of pure platinum about 1.35 in. high and 1.15 in. in diameter. One grain was defined as 1/7,000 part of this pound.

Then, as similarly reported for the unit of length, in 1893 the U.S. Secretary of the Treasury issued an administrative order defining the pound in terms of the international kilogram:

1 lb = 453.592428 g.

As reported for the unit of length, in 1959 the U.S., in concert with the U.K., Canada, Australia, New Zealand, and Africa, agreed on a uniform equivalent for the pound in terms of the kilogram:

1 lb (avoirdupois) = 453.59237 g.

In the U.S., this agreement came from the National Bureau of Standards and the U.S. Coast and Geodetic Survey, with the approval of the Secretary of Commerce. The new relation changed the U.S. equivalent of the pound, in terms of the international kilogram, by 0.1 ppm over what it had been previously.

The international metric unit of mass is the kilogram, kg, which is equal to the mass of the international kilogram maintained at the International Bureau of Weights and Measures at Sevres. The international kilogram is made of a special alloy of platinum with 10% by weight of iridium, and is cylindrical in shape, with approximately the same height and diameter.

Originally, the kilogram was intended to be the mass of 1,000 cm^3 of water at its temperature of maximum density (near 4°C), but, as we have already seen in discussing the liter, there was a difference of 28 ppm.

The United States National Bureau of Standards has two prototypes of the international kilogram, along with several working standard kilograms. It appears that comparisons between two platinum-iridium copies of the international kilogram can be made with an uncertainty of about 0.01 ppm. The international prototype kilograms maintained for the U.S. at the National Bureau of Standards are "Kilogram 4" and Kilogram 20," which were obtained from the International Bureau of Weights and Measures in 1890. Also at the National Bureau of Standards is the "Arago Kilogram" purchased from France in 1821.

D. THE UNIT OF TIME

Of all the natural phenomena observable by man, those occurring in the heavens are the most striking, the most readily observed, and the most regular. It was only natural, then, that in early historical times this regularity was connected with the measurement of time. In the 6th Century B.C., the Ionian Greek philosopher, Thales of Miletus, correctly predicted the time of an eclipse of the sun.

In more modern days, into the 19th Century, the keeping of time for living and working purposes at different locations on the earth's surface has been complicated, with cities and towns maintaining their own individual local or "sun" times. Less than 100 years ago, the railways in Britain ran on London-Greenwich time, while the railways in France ran on Paris time. But in the U.S., then, the great distance from one coast to the other made a difference of several hours in local or "sun" time, so that all the railways in the U.S. needed more than a single time system. Actually, each of the railroads that ran principally north and south, with not much east-west trackage, had its own time. And the long east-west lines, particularly those running from the middle west to the Pacific coast, had several different time zones. This created much confusion at the points where the time systems overlapped.

In 1878, Sandford Fleming, a Scotch-Canadian, proposed the plan of having 24 equal time zones around the earth, each of 1 hr, and each covering 15° of longitude, with London-Greenwich taken as the zero starting point. Railways in the U.S. and Canada adopted the plan, making four time zones in the then continental United States. However, now nearly a century later, a new suggestion is being seriously proposed, arising from the enormous increase in speed of communication and of travel, and the necessity for industry, business, government, and other components of our society to communicate rapidly and freely and transact business at reasonable times. This suggestion is that the continental United States return to one time system. But our existing system of 24 zones around the world is likely to remain with us a long time.

For centuries, the length of the day had been reckoned as the mean time of rotation of the earth on its own axis, with the day split into 24 hr, each hour into 60 min, and each minute into 60 sec, making 1 day equal to 86,400 sec.

Up to 1956, the international unit of time was the second, defined as 1/86,400 part (exactly) of the time required, on the average during a given year, for one complete rotation of the earth on its own axis. But astronomers found that the time of rotation of the earth on its own axis was not quite constant, there being small periodic fluctuations during a given year and small unpredictable changes from one year to another.

It appears that these variations in the time of rotation of the earth on its own axis may be categorized in three ways: secular changes, caused by tidal friction; irregular changes, probably caused by turbulent motion in the liquid core of the earth; and periodic changes, occurring in periods of ½ year caused chiefly by the tidal action of the sun, which slightly distorts the shape of the earth, and in periods of 1 year, caused principally by the seasonal change in the wind patterns of the Northern and Southern Hemispheres.

The secular changes consist of a slow, more or less regular, increase of about 0.0015 sec in a century. The irregular changes come in relatively short periods of time, say 5 to 10 years, with an increase in one period followed by a decrease in the next period. The maximum difference from the mean time for one rotation of the earth on its own axis has been found to be about 0.005 sec during a century. Since 1900 the algebraic accumulation of these irregular differences has amounted to about 40 sec. The periodic changes result in the cumulative effect of the earth being slow in its time of rotation near June 1 of about 0.030 sec and fast in its time of rotation near October 1 of about the same amount. The maximum variation in the length of the day, from one season to another, appears to be about 0.0005 sec.

The secular and irregular variations referred to were discovered by comparing the time of rotation of the earth on its own axis with the time of rotation of the earth and other planets around the sun. The periodic variation was discovered with the aid of quartz crystal clocks. The precision of observation of the periodic and irregular variations given in the foregoing results from the development of the new unit of time discussed later in this section.

Because of the variations described above, the International Committee on Weights and Measures, in 1956, changed the definition of the unit of time from that based on the rotation of the earth on its own axis to one based on the rotation of the earth about the sun. The second was then defined as 1/31,556,925.9747 part of the time required for the earth to orbit the sun in the year 1900. Specifically, the second was taken as the foregoing fraction of the tropical year at 12h, ephemeris time, 0 January, 1900.

One of the difficulties of this definition is the lack of any direct comparison with the second itself. It appears that, while the apparent precision of the foregoing definition* is 1 in 300 billion, the relationship between the definition and the actual realization of the second is of the order of 0.001 per million. This relationship was obtained by a series of astronomical observations over a period of several years.

Meanwhile, spectacular events relating to the measurement of time were taking place, involving the development of atomic beams, masers, and absorption cells for measuring frequency and time. It was found that these newly developed devices could be compared with one another with a precision of better than 0.0001 per million during observations lasting only an hour. Later, the precision was increased significantly.

In 1967, the International General Conference on Weights and Measures approved the following definition of the unit of time, which had been recommended by the International Committee on Weights and Measures in 1964:

> The unit of time is based on the transition between two hyperfine levels ($F = 4, M_F = 0; F = 3, M_F = 0$) of the fundamental state, $2S_{1/2}$, of the atom of the pure nuclide, Cs-133, undisturbed by external fields, with the value 9,192,631,770 cycles (Hertz) taken as (exactly) one second.

Taking the uncertainty as ½ unit in the last figure, this becomes 0.00005 ppm, or the equivalent of 1 sec in 600 years. One of the great advantages of this new unit is that exact calibrations can now be made in a matter of minutes, whereas before enormously long times were required for definitive checks.

It should be noted that the new international unit of time, based on atomic transitions in pure Cesium-133, is uniform and quite independent of the secular, irregular, and periodic variations in the time of rotation of the earth on its own axis, referred to previously. The time registered by the "atomic" clock can be adjusted to accord with mean solar calendar time given by the rotation of the earth.

In the foregoing discussion, the word second has uniformly meant the "mean solar second," which is to be distinguished from the "sidereal second" of the astronomer. The relation between the two is as follows:

1 sidereal sec = 0.9972696 sec.

For the benefit of users everywhere, in the laboratory, in industry, in the marketplace, and in the home, the U.S. Government is providing time and frequency services 24 hr a day from several radio stations operated by the National Bureau of Standards and by the U.S. Department of the Navy.

The U.S. Navy has ten different radio stations (NBA, NSS, NLK, NAA, NPM, NWC, NPN, NPG, NDT, and Omega) which broadcast time and frequency.[10]

The National Bureau of Standards has two radio stations which broadcast continuously day and night, WWV at Fort Collins, Colorado, and WWVH at Mauai, Hawaii.[7] The services provided include the following: (a) standard radio frequencies of 2.5, 5, 10, 15, 20, and 25 MHz (10^6 cycles); (b) standard time voice announcements, each minute; (c) standard time intervals of 1 sec and 1 min; (d) corrections to adjust mean solar time to astronomical time; and (e) standard audio frequencies of 440, 500, and 600 Hz.

These radio signals of the National Bureau of Standards are controlled against the new international unit of time with an accuracy near 0.001 ppm. These transmissions of time and frequency are coordinated through the International Bureau of Time in Paris in accord with international agreements, and are based on the international time scale, Universal Coordinated Time (UTC), more commonly known as Greenwich Mean Time. Prior to January 1, 1972, the NBS time signals were kept in close agreement with "astronomical time," but beginning at that date this was discontinued. The UTC maintained by the National Bureau of Standards is no longer adjusted periodically to agree with the rate of rotation of the earth, and gains about 1 sec per year on "earth rotation time." Corrections to UTC are now made in step adjustments of exactly 1 sec as directed by the International Bureau of Time. These "leap" second adjustments ensure that UTC signals as

*As explained by McNish,[3] this "multidigited number was obtained from Simon Newcomb's equation for the celestial motion of the sun. The equation is quadratic in time, and gives, subject to correction for periodic effects, the longitude of the sun in the plane of the ecliptic with respect to the vernal equinox. The particular time in the definition reduces the quadratic term in the equation to zero. This is the ephemeris second, the unit of time in terms of which all planetary motions were most simply expressed."

broadcast will not differ from "earth rotation time" by more than ±0.7 sec. These corrections no longer relate to "astronomical time." The "leap" second adjustments are made at the end of the UTC month, preferably on December 31 or June 30. Two "leap" seconds were inserted in 1972, one on June 30, 1972, and one on December 31, 1972. Appropriate insertions followed in 1973. At midnight on December 31, 1973, one "leap" second was put in.

E. DISCUSSION

Standards of measurement of length, mass, and time are important not only to the scientist, engineer, industrialist, and businessman, but also to private individuals in their daily lives, including the purchase of necessities for their living – food, clothing, heat, energy, etc.

An example of the manner in which the international standards flow down to everyday life in the U.S. may be seen from the standard of mass. First is the "International Kilogram" maintained at the International Bureau of Weights and Measures at Sevres, near Paris. Then we have "Kilogram 20," the national prototype kilogram of the U.S. maintained at the National Bureau of Standards, which was calibrated at the International Bureau of Weights and Measures. Then we have National Bureau of Standards Laboratory Standards (both metric and customary–English) calibrated at the National Bureau of Standards. Next we have State Reference Standards which have been calibrated at the National Bureau of Standards. Then we have State Laboratory Standards calibrated in the State Laboratories against the State Reference Standards. Then we have State Field Standards and County and City Standards which serve to provide regulatory controls in the marketplace.

In the case of the unit of length, the interrelations are the same except in the initial step, where reference to the "International Standard Meter Bar" at Sevres is no longer necessary, since the national prototype for the United States is now calibrated against the fundamental unit of length as determined by specified monochromatic radiation from pure Krypton-86.

In the case of the unit of time, the flow to the marketplace and the consumer is quite direct since the National Bureau of Standards broadcasts accurate time and frequency signals over its radio stations in Colorado and Hawaii, as described in the preceding section.

As indicated earlier in this chapter, the legal standards of length, mass, and time in the U.S. are as follows:

Length (new definition, 1960) – The unit of length is the meter, m, which is equal exactly to 1,650,763.73 wavelengths of light in vacuo produced by the unperturbed transition $2p_{10}$ - $5d_5$ in the pure nuclide Krypton-86. (This is the orange-red radiation of Krypton.)

Mass (legalized in 1866, present U.S. standard received in 1890) – The unit of mass is the kilogram, kg, which is equal to the mass of the international kilogram maintained at the International Bureau of Weights and Measures.

Time (new definition, 1964) – The unit of time is based on the transition between the two hyperfine levels of the ground state of the atom of the pure nuclide Cesium-133, undisturbed by external fields, with the value 9,192,631,770 cycles taken as exactly 1 sec.

It should be noted that of the foregoing three standards, only one is an artifact, that of mass, the other two, of length and time, being now fully and completely reproducible anywhere at any time in any laboratory having the appropriate apparatus and equipment. No longer need we be concerned about the security and validity of the International Meter Bar in the vaults of the International Bureau of Weights and Measures. However, our concern must continue with respect to the security and validity of the International Kilogram.

There are, in principle, several ways by which the unit of mass could be maintained apart from the International Kilogram. The difficulty here is that the present precision of comparison of kilogram masses, with reference to the International Kilogram, is of the order of 0.01 ppm. And the precision attainable with these alternate methods is several magnitudes less. Until a precision of 0.001 ppm or better is attainable by any new method, it is likely that we will continue with the "International Kilogram."

Following is one of the methods that could conceivably provide a unit of mass. The Josephson Effect, discovered in 1962, is produced by a "weak" spot in a superconducting system, as illustrated in Figure 2.2 where the "weak" spot is identified by the cross-hatched area. When a voltage, V is applied at the ends of the "weak"

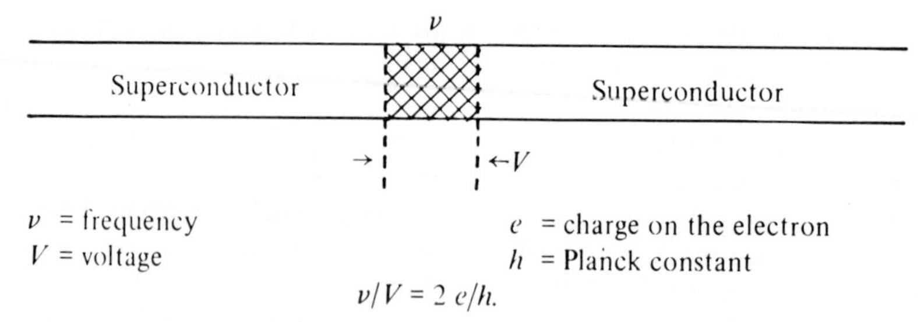

FIGURE 2.2. Josephson Effect. A weak spot, indicated by the cross-hatching, in a superconducting system at which a localized quantum phase difference can be established between two superconducting regions and monitored by noting the frequency as a function of the direct current voltage applied.

spot, a frequency, ν, develops which can be measured accurately. The relation involved is

$$\nu/V = 2e/h, \text{ or} \quad (1)$$

$$V = \nu h/2e \quad (2)$$

Now the voltage, V, and the frequency, ν, can be measured quite accurately. Reliable values for the Planck constant, h, and the electronic charge, e, are available from other measurements. Now the voltage, V, can be replaced by $A \cdot \Omega$, the product of the current, A, and the resistance, Ω. The current can be measured by a force, which can be set equal to the acceleration times mass. In this rather "longwinded" way, one can arrive at a unit of mass. The only difficulty with this procedure is that the resulting precision of measurement will be far short of the desired 0.001 ppm.

Another method for arriving at a unit of mass independent of the International Kilogram is the following. Consider the Avogadro constant, N, which is the number of molecules (or atoms or ions, etc.) in one gram-formula weight of any given species. One can determine the value of the Avogadro constant as follows:

1. Select a "perfect" crystal of a pure chemical substance, say $CaCO_3$
2. Calculate the molecular weight, M of $CaCO_3$
3. Measure the mass, m of the given crystal
4. Measure the volume, V of the given crystal
5. Make appropriate X-ray measurements of the given crystal so that the angles and spacings are known, permitting calculation of the volume, ν, which can be assigned to one molecule of the substance, $CaCO_3$.

Then the following relations apply:

$$N = (V/\nu)/(m/M) \text{ molecules/mole} \quad (3)$$

But

$$N = \mathfrak{F}/e, \quad (4)$$

where $\mathfrak{F}$ is the Faraday constant and e is the electronic charge. Substituting, and rearranging, one obtains:

$$m = VMe/\mathfrak{F}\nu \quad (5)$$

Each of the quantities on the right side can be determined by experiment, independent of the unit of mass, and hence provides a method for determining the unit of mass. As before, the difficulty here is the fact that the overall precision attainable is far less than the desired 0.001 ppm.

Perhaps, at some future time, someone may develop a sufficiently precise and accurate method for establishing a unit of mass independent of any artifact.

It may be well here to report that, at its meeting in October, 1973, the International Committee on Weights and Measures adopted several recommendations submitted by its Advisory Committee on the Definition of the Meter, as follows:[11]

1. That the value $3{,}392{,}231.40 \times 10^{-12}$ m be used for the wavelength emitted by a helium-neon laser stabilized to the P (7) line in the ν 3 band of the methane molecule.

2. That the value $632{,}991.399 \times 10^{-12}$ m be used for the wavelength emitted by a helium-neon laser stabilized to the "i" component of the R(127) line in the 11-5 band of Iodine-127.

3. That the value 299,792,458 m s^{-1} be used for the velocity of light in vacuum.

The Committee considers the uncertainty in the three foregoing numbers to be 0.004 ppm. Carried to its logical conclusion, the foregoing action will lead to a new definition of the unit of length in terms of the second and the velocity of light in vacuum.

For additional information on this subject, the reader is referred to the references and to the headquarters of the organizations mentioned in the text. The author is indebted for special information and suggestions to several friends and former associates of the National Bureau of Standards (U.S.A.): Allen V. Astin, former Director; Richard D. Deslattes, Chief of the Quantum Metrology Section; and Alvin G. McNish, former chief of the Metrology Division.

F. REFERENCES

1. **Guillaume, C.-E. and Volet, C.**, National and local systems of weights and measures, in *International Critical Tables of Numerical Data, Physics, Chemistry, and Technology,* Vol. 1, Washburn, E. W., Ed., McGraw-Hill, New York, 1926, 1.
2. Measures and weights; Metrology, in Vol. 15; Time measurement, in Vol. 22, in *Encyclopaedia Britannica,* 14th ed., Encyclopaedia Britannica, New York, 1932.
3. **McNish, A. G.,** *Fundamentals of Measurement,* Electro-Technology-Science and Engineering Series No. 53, Electro-Technology, New York, 1963.
4. **McNish, A. G.,** Dimensions, units and standards, *Phys. Today,* 10, 19, 1957.
5. **McNish, A. G.,** The basis of our measuring systems, *Proc. Inst. Radio Eng.,* 47, 636, 1959.
6. **De Simone, D. V.,** *A Metric America,* National Bureau of Standards Special Publication 345, U.S. Government Printing Office, Washington, D.C., 1971.
7. **Viezbicke, P. P.,** *NBS Frequency and Time Broadcast Services,* National Bureau of Standards Special Publication 236, U.S. Government Printing Office, Washington, D.C., 1973.
8. **Judson, L. V.,** *Weights and Measures of the United States. A Brief History,* National Bureau of Standards Miscellaneous Publication 247, U.S. Government Printing Office, Washington, D.C., 1963.
9. **Page, C. H. and Vigoureux, P.,** *The International System of Units (SI),* National Bureau of Standards Special Publication 330, U.S. Government Printing Office, Washington, D.C., 1972.
10. Time Service Publications, U.S. Naval Observatory, Washington, D.C. 20390.
11. Proceedings of the International Committee on Weights and Measures, International Bureau of Weights and Measures, Sevres, France, October 1973.

Chapter 3

THE INTERNATIONAL METRIC SYSTEM

A. INTRODUCTORY COMMENTS

Units of measurement are a necessary part of living and working for all people in the world. In the early days, when countries were more or less isolated from one another, each country developed its own set of measures. Then, as the intellectual status of the people of the world advanced in science and technology, improvements came in the accuracy and precision of measurement. At the same time there came an enormous increase in the speed of communication and travel. This had the effect of bringing all the peoples of the world closer together and increasing greatly trade among them. This meeting of the different peoples in the marketplace brought forth the need for a common world system of measurement.

In this chapter we review the following: the development of the metric system in France beginning almost two centuries ago; the adoption of the metric system by many other countries; the manner in which conversion to the metric system was accomplished in Japan and the United Kingdom, two countries which committed themselves to the metric system relatively recently; the history of the metric system in the United States; the present status of the practical conversion to the metric system by the U.S.; and, finally, a discussion of the new International Metric System (SI).

B. DEVELOPMENT OF THE METRIC SYSTEM IN FRANCE

As early as the 17th Century, the placing of the system of weights and measures on a scientific basis had been proposed by Jean Picard, the astronomer. He suggested using as the unit of length the length of a pendulum beating 1 sec at sea level and at a latitude of 45°.

The French Revolution of 1789 gave some impetus to the movement toward a new system of measurement. The French National Assembly in 1790 appointed a committee of the French Institute for the Sciences and the Arts to consider the suitability of adopting (a) the length of the seconds pendulum, (b) a fraction of the length of the equator, or (c) a fraction of the quadrant of the terrestrial meridian. The committee's report, made in 1795, decided in favor of using one ten-millionth part of the quadrant of the earth's meridian passing through the poles and intersecting the equator at the earth's surface. Some provisional standards were made available. Then another committee was appointed in 1795 to draw up a system of weights and measures based on the meter and the nomenclature to accompany it. As part of this work, a survey was actually made of a section of the meridian between Dunkirk, France, and Barcelona, Spain, carried out under rather dangerous conditions arising from the French Revolution. The results were implemented by making an appropriate platinum bar, as an "end" standard. This committee made its report on the length of the meter in 1799. This report resulted in action by the French National Assembly.

The law of December 10, 1799, fixed the value of the meter and the kilogram, and decreed the new metric system compulsory beginning in 1801. About 1800, France sponsored an international conference in Paris, the purpose of which was to inform other countries about the new metric system and to exhibit the new standards.

Because commercial and household weights and measures of the new system were scarce, use of the legal metric system was not enforced. Popular acceptance therefore came very slowly. In 1812, Napoleon Bonaparte, recognizing the practicality of having adequate supplies of measures under the new system and that time would be required to do this, issued a decree that partially reinstated the old system while continuing the metric system as the standard.

In 1837 the French National Assembly finally and officially restored the metric system in France and legislated that its use should be compulsory throughout the country beginning January 1, 1840. Following this definitive action by France, the metric system began to spread rapidly to other countries.

C. ADOPTION OF THE METRIC SYSTEM BY OTHER COUNTRIES

As mentioned above, the metric system did not spread to other countries, except those bordering on France, until after France had definitely established the metric system in its own country. Then the metric system began to take hold rapidly in other countries.

Following is a tabulated summary of the number of countries definitely committed to the metric system by years, taken from a chart published by the Metric Association[3] in the U.S.: by 1840, 4 countries; 1860, 9 countries; 1880, 25 countries; 1900, 37 countries; 1920, 54 countries; 1940, 68 countries; 1960, 78 countries; 1970, 83 countries. Figure 3.1 shows graphically the advance of the metric system in the world.

As of 1971, the report by DiSimone[2] of the National Bureau of Standards indicated that the only large country of the world not then committed to the metric system was the United States. In company with the U.S. as uncommitted to the metric system then were the following twelve countries:

Barbados	Jamaica	Sierra Leone
Burma	Liberia	Southern Yemen
Gambia	Muscat and Oman	Tonga
Ghana	Nauru	Trinidad

Different countries have converted to the metric system in different ways. The inconvenience and cost of conversion depends greatly on the scientific and technological development of a given country. The greater the scientific and technological status of a country, the greater the inconvenience and the cost. Also, the more rapid the conversion, the greater the inconvenience and the cost. The following two sections summarize the conversion to the metric system in two highly industrialized countries which have committed themselves to the metric system in recent years – Japan and the United Kingdom.

D. CONVERSION TO THE METRIC SYSTEM IN JAPAN

The national conversion to the metric system in Japan was accomplished in an indirect manner.

Before 1891, there was a great diversity of units of measure in Japan. About 1891, the Japanese units of measure were defined in terms of metric equivalents, much as was done in the United States. The principal unit of length was the shaku, defined as follows:

1 shaku = 10/33m.

Multiples of the shaku included the following: shi, 10^{-5}; mô, 10^{-4}; rin, 10^{-3}; bu, 10^{-2}; sun, 10^{-1}; yabiki, 2.5; hiro, 5; ken, 6; jô, 10; chô, 360.

The principal unit of mass was the kwan, defined as follows:

1 kwan = 15/4 kg.

Multiples of the kwan included the following: shi, 10^{-7}; mô, 10^{-6}; rin, 10^{-5}; fun, 10^{-4}; mommé, 10^{-3}; niyo, 0.004; hyaku-mé, 0.10; kin, 0.16; ninsoku-ichi-nin, 7; hiyak-kin, 16; karushiri-ichi-da, 18; kommo-ichi-da, 40.

Area for land measure was not related to the unit of length in a simple manner. The unit for this purpose was the bu, defined as follows:

1 bu = 100/30.25 m².

Multiples of this unit for land area included the following: gô, 0.1; tsubo, 1; sé, 30; tan, 300; chô, 3,000.

By 1921, Japan had three officially recognized systems of weights and measures: the traditional Japanese system, defined in terms of metric equivalents; the English system; and the metric system. Then, in 1921, the Japanese Diet passed a law extending the use of the metric system, indicating preference for it over the other two systems. Also, instruction in the metric system was introduced into the primary schools of Japan at that time. Plans were laid for a complete conversion to the metric system over a period of 10 years for government agencies, public utilities, and a few industries, while other sectors of the economy were given 20 years for the conversion. But, unfortunately, progress was slow and the time schedules were lengthened to 15 and 30 years, respectively.

Then, in 1939, near the beginning of World War II, a new law of the Japanese Diet restored the traditional Japanese system, diffused in terms of metric equivalents, to equal footing with the metric system. At the same time, this law extended to 1958 the final conversion to the metric system.

Following the end of World War II in 1945, during the occupation of Japan by the U.S., the units of measurement used in the U.S. came into prominent use in Japan. Then, in 1951, another law passed by the Japanese Diet affirmed the year 1958 as the date for final conversion to the metric

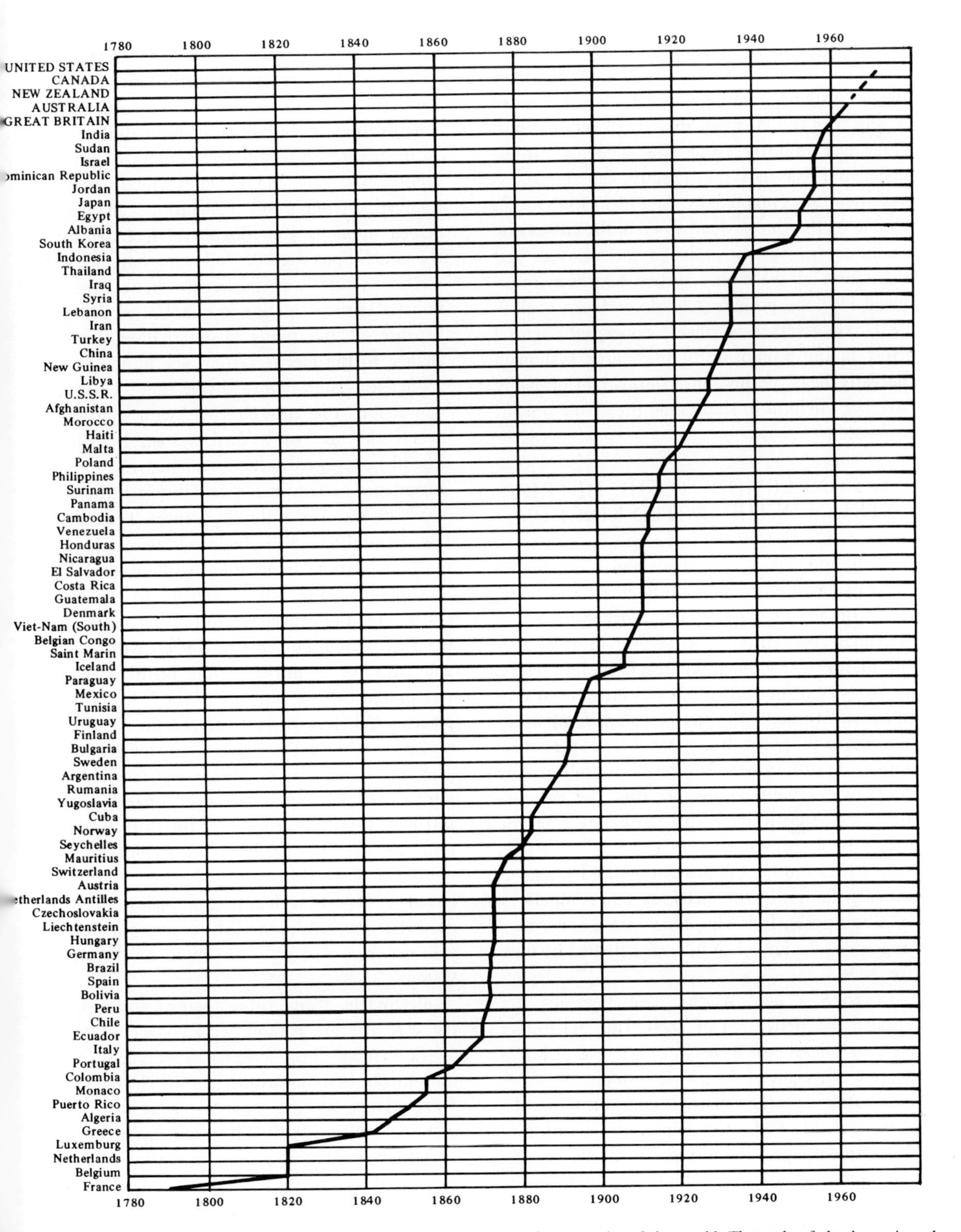

IGURE 3.1 Chart showing the advance of the metric system among the countries of the world. The scale of abscissae gives the ears. (From Helgren, F. J., *Metric Units of Measure,* 7th ed., Metric Association, Waukegan, Ill., 1967. With permission.)

system. Most of the planning for the conversion was done by a committee on Promotion of the Metric System, a semipublic body, which worked closely with the Japanese Ministry of International Trade and Industry. While the schedule for the changeover was not met exactly, the conversion was essentially complete by the early 1960's.

One of the things that helped in the conversion to the metric system in Japan was the fact that instruction in the metric system had been introduced into the primary schools of Japan back in 1921. However, a stronger, concerted, central effort in the promotion of the metric system in the early days of the conversion would probably have been helpful in shortening the actual time of conversion.

E. CONVERSION TO THE METRIC SYSTEM IN THE UNITED KINGDOM

In Britain, many individuals and a number of trade associations used their influence to make the use of the metric system compulsory in Britain. A "Decimal Association" was formed in 1854, but did not make very much progress in promoting the metric system. In 1864 a bill was introduced into Parliament to make the metric system compulsory for certain purposes, but there were some objections from government sources and a bill making the metric system permissible (but not compulsory) was substituted. This latter became the Metric Act of 1864.

In 1871, another bill was put before Parliament to make the use of the metric system compulsory for all purposes, following a high-speed conversion program of 2 years. This bill was defeated by only five votes.

In 1878 a new Weights and Measures Act repealed the Metric Act of 1864. This Act in fact made it illegal to have in one's possession a measure of weight of the metric system.

Debates on the matter continued in the Parliament. In 1893 a representative delegation of businessmen urged the British Chancellor of the Exchequer to push for adoption of the metric system, but he did not.

In 1897 Parliament passed the Weights and Measures (Metric System) Act, which made legal, but not compulsory, the use of the metric system in trade. This Act thereby abolished the penalty for having a weight or measure of the metric system in one's possession. In 1907 another effort was made in Parliament to make use of the metric system compulsory, but this effort was unsuccessful. Nothing substantial was done toward "metrication" in Britain until about 1950.

In that year a government committee released a report on weights and measures which contained a number of important conclusions unanimously agreed upon by the committee. Included among the conclusions were the following statements: the metric system was inherently better than the imperial system in use in the United Kingdom; because of the needs in international trade, a change to the metric system was bound to come sooner or later; for a given time schedule of conversion, the cost increases with every year's delay; the use of a dual system (metric and English) causes extra inconvenience; the long-term advantages of conversion far outweigh the inconvenience and cost of the changeover. The report of this committee also included two additional recommendations in addition to the principal one of supporting the full conversion to the metric system: (a) the conversion to the metric system in the U.K. should be made in concert with the U.S. and with the countries of the British Commonwealth and (b) the currency of the U.K. should be placed on a decimal basis before the full conversion of weights and measures to the metric system. The report of this committee was received with interest but no impetus for the conversion came forth. In 1960 a committee of the British Association for the Advancement of Science and the Association of British Chambers of Commerce issued a report saying that a majority of the industrial concerns in the U.K. considered the time still not ripe for full conversion to the metric system.

Meanwhile, many other countries with which the U.K. did much trade were converting to the metric system. These included several countries of the British Commonwealth. In 1963 the British Standards Institution made a survey and found that a large majority of industrial firms in the U.K. were then in favor of proceeding without delay to a full conversion to the metric system without waiting for the U.S. and the other countries of the British Commonwealth not yet committed to the metric system.

In 1965 the President of the Federation of British Industries informed the British Government's Ministry of Technology that a majority of its firms favored adoption of the metric system as the primary system, and, ultimately, with full conversion, as the only system of measurement to be used, and requested the government to support such conversion and to aid in its scheduling. The reply of the Ministry was favorable and indicated that much of the burden of conversion must be born by British industries, sector by sector, and that the government would assist the program by using metric specifications in its purchases at appropriate times and on appropriate materials in the process of conversion. It was on May 24, 1965, that the President of the British Board of Trade announced that the U.K. would convert to the metric system over the course of the following 10 years.

In 1968, following 2 years of study of the problem, the Ministry of Technology issued a report which had three principal points: (a) manufacturing industry can make the change efficiently and economically only if the economy as a whole moves in the same direction on a broadly similar time scale, and in an orderly way; (b) a Metrication Board should be established to guide, stimulate, and coordinate the planning for the conversion; and (c) any legal barrier to the use of the metric system should be removed – that is, tariff and other regulations written in terms of the English system of units should be rewritten. It was further stated that (a) every sector of the economy need not move at the same pace; (b) the costs of converting to the metric system must be where they fall; (c) the year 1975 was taken as the target date for conversion, with the understanding that some sectors of the economy would aim at an earlier date and some at a later date.

The Metrication Board, an advisory body, had representation from industry, business, education, and the public. It is interesting to note the time schedule followed by various sectors of industry in the U.K. for the conversion to the metric system, starting with the first concrete moves as of January 1, 1969, following several years of planning (in the following listing, the date indicates the year after which only the metric system would be available in the given sector): construction industry, 1972; electrical industry, 1969–1975; marine industry, 1970–1972; engineering industry, 1969–1975; commodity industry, 1970–1972.

In the U.K., the educational system has been specially helpful in the changeover. Publishers and manufacturers of educational material have co-operated in the plan. The several industries, through the appropriate associations, provided guidance through training schools.

It appears that the conversion to the metric system in the U.K. has been proceeding substantially on schedule.

F. HISTORY OF THE METRIC SYSTEM IN THE UNITED STATES

In our discussion in Chapter 2 of the units of length and mass in the United States, we made mention of the metric system a number of times. In fact, the U.S. was not too far from adopting the metric system as early as 150 years ago. Actually, the metric system was made legal, but not compulsory, in the U.S. in 1866, at which time also our customary English units were defined in terms of the metric units. But, in spite of many attempts, full conversion to the metric system has not yet come about, although, as is indicated in the next section, we are on the threshold of full conversion. Following is a summary of the history of the metric system in the United States.

In 1817 the U.S. Coast Survey obtained its "Committee Meter (Iron)," which was one of 12 copies of the first platinum metric standard. This meter, which had originally been brought to the U.S. in 1805, was used by the U.S. Coast Survey until 1890, when the U.S. officially obtained its metric standards (see below).

In 1821 the U.S. obtained by purchase from France its "Arago Platinum Meter," now in the custody of the National Bureau of Standards.

In 1821 the then Secretary of State and later President of the United States, John Quincy Adams, submitted a report to the U.S. Congress entitled "Report upon Weights and Measures." This report included the first official systematic consideration of the metric system by the U.S. Government. The report called attention to the advantages of the metric system: (a) the simplicity of the unit of length and of the unit of mass; (b) the use of decimal multiples of length and mass; (c) the derivation of units of area and volume from the unit of length; and (d) the uniform and precise terminology of the system.

As reported in Chapter 2, Adams' report suggested four possible paths for the U.S. to follow: (a) to adopt the new French metric system; (b) to restore the old English system; (c) to devise and establish a new combined system;

and (d) to continue with the then current U.S. system, merely fixing the standards. But the report recommended that the U.S. should (a) proceed to standardize the customary English units in use and (b) later begin negotiations with France, Britain, and Spain to establish a uniform international system of measurement. These latter recommendations were practical in the sense that (a) by 1821 many states of the U.S. had already enacted laws specifying the familiar English units, and contrary action by the Federal Government would involve the problem of States Rights, and (b) that most of the foreign trade of the U. S. at that time was with Britain, and the U.S. was bounded by Canada and by Spanish possessions. But the U.S. Congress took no action on the report of Adams.

In 1832 the U.S. Department of the Treasury adopted English standards to meet the needs of the customs houses.

In 1863, President Lincoln approved the formation of the National Academy of Sciences to advise the government on all technical matters. At the request of the U.S. Secretary of the Treasury, a committee of the National Academy of Sciences was formed, under the chairmanship of the eminent physicist, Joseph Henry, to consider the problem of weights, measures, and coinage. In 1865 this committee submitted its report, recommending adoption of the metric system.

In 1866 the newly appointed Committee on Coinage, Weights, and Measures of the U.S. House of Representatives, under the chairmanship of Congressman John A. Kasson, reported three bills dealing with the metric system, which were passed by the Congress. The first of these made legal, but not compulsory, the use of the metric system in the U.S. and specified the customary English units in terms of their metric equivalents. The second bill directed the U.S. Postmaster General to distribute metric postal scales to all post offices handling foreign mail. The third bill directed the U.S. Secretary of the Treasury to furnish each state with one set of metric standards. It is important to repeat here the wording of the law enacted by the U.S. Congress in 1866, making the metric system legal, but not compulsory:

> It shall be lawful throughout the United States of America to employ the weights and measures of the metric system; and no contract or dealing, or pleading in any court, shall be deemed invalid or liable to objection because the weights and measures expressed or referred to therein are weights and measures of the metric system.

However, the practical conversion of the U.S. economy to the metric system was held back because of the opposition, not only of many ill-informed persons but also of many others, including prominent educators, who raised numerous objections.

In 1873 Frederick A. P. Barnard, president of Columbia College in New York, who had been a staunch supporter of the metric system, founded and became president of the American Metrological Society, created to promote conversion to the metric system. This Society spawned the American Metric Bureau, with Barnard as president and Melvil Dewey, who later developed the Dewey classification system for libraries, as executive director. The American Metric Bureau engaged in many practices designed to support conversion to the metric system, including the provision, at low cost, of metric measures to educational institutions.

In 1875 the U.S. was one of 17 countries that signed the Convention or Treaty of the Meter. This convention did the following: (a) reformulated the metric system and refined the accuracy of its standards; (b) provided for the construction of new measurement standards; (c) provided for the distribution of copies of the new standards to participating countries; (c) established the International Bureau of Weights and Measures as a world repository and laboratory, at Sevres, near Paris, on a piece of international territory donated by France; and (e) made arrangements for holding international conferences, supported by the work of appropriate international committees, to further international action and agreements in the future.

Between 1877 and 1886, several pieces of legislation dealing with the metric system came before the U.S. Congress, but nothing definite resulted.

In 1879, in Boston, Massachusetts, the International Institute for Preserving and Perfecting Weights and Measures was founded. This Institute made quite clear that the weights and measures to be preserved were those of the English system. As reported by De Simone:[2]

> The International Institute's thinking was greatly influenced by a contemporary movement known as pyramidology. The main contention was that the Great Pyramid at Giza, Egypt, had been constructed by the hand of God in such a way that it contained all of His scientific gifts to mankind. By elaborately manipulating the pyramid's dimensions, pyramidologists proved that the Anglo-Saxon

race was one of the lost tribes of Israel and that Anglo-Saxon weights and measures, represented by the customary English system, were of divine origin. The Institute was naturally opposed to any other measurement system and even wanted to purify the English system by eliminating all non-Anglo-Saxon influences.

One of the principal targets of this Institute was the fact of the adherence of the U.S. to the Convention of the Meter.

In 1889 the new metric standards provided for by the 1875 International Convention of the Meter were completed and the U.S. received its prototype copies of the International Meter and the International Kilogram.

In 1893 the U.S. Secretary of the Treasury issued an administrative order (prepared by the superintendent of its Office of Weights and Measures, which was the official custodium of weights and measures for the U.S. prior to the founding of the National Bureau of Standards in 1901) declaring the new metric standards to be the fundamental standards of length and mass for the U.S.

In 1896 Congressman Dennis Hurley introduced a bill in the U.S. House of Representatives which provided that all departments of the U.S. Government should employ and use only the weights and measures of the metric system in transacting official business and that, 3 years later, in 1899, the metric system would become the only legal system of measurement recognized in the U.S. This bill, strongly supported by the U.S. House of Representatives Committee on Coinage, Weights, and Measures, passed the House by the narrow margin of two votes, 119 to 117. But, as reported by De Simone,[2] opponents of the bill immediately forced reconsideration and the bill was returned to the Committee, where it died. The reasons given for the demise of this bill included the failure to brief other Congressmen fully on the matter, and the worry of Congressmen about adverse reaction to the bill from farmers and tradesmen in an election year.

From 1897 to 1907, a number of bills were introduced in the U.S. Congress dealing with the metric system. Most of these bills provided for adoption without delay of the metric system by the U.S. Government in its official transactions and for later adoption by the other sectors of the economy. In this period the proponents reiterated the usual advantages of the metric system, plus the fact that more and more countries, worldwide, were adopting the metric system (37 countries by 1900): (a) the conversion would have to be made eventually; (b) the conversion becomes more difficult and more costly the longer the delay; (c) the intrinsic simplicity and utility of the metric units were unmatched by any other system; and (d) the decimalization of the metric system fitted in with our system of coinage and made it easier to learn.

According to De Simone,[2] "The opposition (to the metric system) was better organized and more effectively led than ever before. It was spearheaded by two men . . . . They rallied the support of engineers, manufacturers, and workmen, and claimed to be practical men, not closet philosophers or theorists. They charged that the metric system had been a practical failure in countries which had adopted it."

From 1907 until the early 1930's, opposition to the metric system flourished. The American Institute of Weights and Measures was established, with backing from large manufacturing firms and associations. In addition to publishing its own pamphlets, this Institute received support from some leading professional and trade journals. By this time the main arguments against the metric system were being embellished with rather outlandish statements, as evidenced by a series of articles appearing in 1920 under titles such as the following: "What Real He-Men Think of the Metric System;" "Metric Chaos in Daily Life;" "A Metric Nightmare."

In 1916 the American Metric Association was established in New York, and in 1917 the World Trade Club was established in San Francisco. Both of these organizations were set up to support conversion to the metric system and began issuing important statements and reports in favor of it. The American Metric Association had the support of scientists, educators, engineers, physicians, pharmacists, etc.

Although many bills relating to the metric system were introduced in the U.S. Congress, nothing significant further occurred through the post World War I period, the depression of the 1930's, World War II, and a few years after World War II.

In 1957, when the "space age" was launched by the U.S.S.R. with its first satellite, things began to change rapidly regarding the metric system. The U.S. Department of the Army established the metric system as the basic system of measurement for weapons and related equipment. A Committee of the Organization of American States proposed adoption of the metric system in all countries of North and South America.

In 1958 the U.S., the U.K., Canada, Australia, New Zealand, and South Africa, in which countries the metric system was then legal but not compulsory, agreed to the same metric equivalents to define the inch and the pound.

In 1959 three bills were introduced into the U.S. Congress to deal with conversion to the metric system. Two of these specified that a study be undertaken to investigate conversion to the metric system, and the third requested the President to take steps toward the total adoption of the metric system. None of these bills was enacted. Meanwhile, word came in May, 1965, that the U.K. had committed itself to full conversion to the metric system.

Finally, in August, 1968, the Congress passed Public Law 90-472, "to authorize the Secretary of Commerce to make a study to determine advantages and disadvantages of increased use of the metric system in the United States."

G. PRESENT STATUS OF THE PRACTICAL CONVERSION TO THE METRIC SYSTEM IN THE UNITED STATES

As indicated above, the U.S. Congress enacted Public Law 90-472 on August 9, 1968, authorizing the U.S. Secretary of Commerce to proceed with a study of total conversion to the metric system in the U.S., and assess advantages, disadvantages, costs, savings, indirect benefits, etc. The Act required the Secretary of Commerce to submit a final report within 3 years of the date of the Act.

In July, 1971, the U.S. Secretary of Commerce submitted to the Congress a report entitled "A Metric America – A Decision Whose Time Has Come," prepared at the National Bureau of Standards under the direction of D. V. De Simone and published as National Bureau of Standards Special Publication 345.[2] This report carries the following recommendations from the Secretary of Commerce:

1. That the United States change to the International Metric System deliberately and carefully
2. That this be done through a coordinated national program
3. That the Congress assign the responsibility for guiding the change to a central coordinating body responsive to all sectors of our society
4. That, within this guiding framework, detailed plans and timetables be worked out by these sectors themselves
5. That early priority be given to educating every American schoolchild and the public at large to think in metric terms
6. That immediate steps be taken by the Congress to foster U.S. participation in international standards activities
7. That, in order to encourage efficiency and minimize the overall costs to society, the general rule should be that any changeover costs shall "be where they fall"
8. That the Congress, after deciding on a plan for the nation, establish a target date 10 years ahead, by which time the U.S. will have become predominantly, though not exclusively, metric
9. That there be a firm government commitment to this goal.

The main report itself contains 190 pages and is supported by 12 supplemental reports giving considerable detail on various aspects of the study. The chapters of the main report have the following headings: Perspective; Two Centuries of Debate; Measurement Systems; Arguments That Have Been Made for Metric and for Customary (Units); Going Metric – What Would It Really Mean?; The Metric Question in the Context of the Future World; Going Metric – the Broad Consensus; Recommendations and Problems Needing Early Attention; Benefits and Costs; Two Paths to Metric – Britain and Japan.

The study called for by Congress had been assigned by the Secretary of Commerce to the National Bureau of Standards. The Secretary of Commerce appointed a Metric System Study Advisory Panel consisting of 50 members from organizations representing a wide variety of interests. The function of this Panel was to participate in the planning and conduct of the study and to help ensure that an opportunity was provided for all sectors of society to be heard. The plan for the study was completed in December, 1969, and provided for a series of hearings, called National

Metric Study Conferences, supplemented by a number of special investigations.

Seven National Metric Study Conferences were held during the late summer and fall of 1970, in the following categories: (a) labor; (b) consumer affairs; (c) education; (d) construction; (e) engineering-oriented industry; (f) consumer-related industry; (g) small businesses, state and local governments, natural resources, health, transportation, and other services. Contributions were invited from more than 700 groups in these categories, which included labor unions, trade associations, professional societies, educational associations, consumer-related organizations, etc.

The investigations that supplemented the National Metric Study Conferences covered the following subjects: (a) manufacturing industry; (b) nonmanufacturing business; (c) education; (d) consumers; (e) international trade; (f) engineering standards; (g) international standards; (h) Department of Defense; (i) Federal civilian agencies; (j) commercial weights and measures; (k) history of the metric system controversy in the U.S.

The recommendation of the report to achieve a substantial conversion to the metric system in the U.S. by plan within a period of 10 years permits many costs to be phased out in simple replacement of parts, thus minimizing, in this respect, the cost of the changeover. More rapid conversion by plan results in increased costs of this kind. If no plan is used, estimates indicate that a substantial changeover might take place in about 50 years, but the balance between benefits and costs in such a case becomes very much less favorable. Among the many kinds of benefits resulting from conversion are the following: calculations are simplified in all sectors of society; compatibility, interchangeability, and repair of apparatus and equipment are greatly improved; significant reduction of the number of sizes of equipment and parts can be achieved; foreign trade is increased; domestic business is facilitated; scientific, technological, and industrial communication is facilitated. As of April, 1974, there were a number of bills before the U.S. House of Representatives dealing with implementation of the recommendations of the Metric Study reported above. It is expected that the U.S. Congress will take appropriate action and soon authorize the planned conversion to the metric system in the U.S.

H. THE INTERNATIONAL METRIC SYSTEM (SI)

The present International Metric System, or, in the original French, Système International d'Unites (abbreviated as SI), is based on the original metric system involving the meter, kilogram, and second. The "cgs," or centimeter-gram-second, system used, as the name indicates, the centimeter and the gram as the base units of length and mass. The later "MKS" system used the meter and the kilogram, along with the second, as the base units. The International Metric System (SI) was a refinement of the metric system ("MKS" version) to satisfy the highest possible scientific requirements among the countries of the world, and, in addition, used the meter and the kilogram rather than the centimeter and the gram as the base units. The main reason for this change was to facilitate transfer of the units of length and mass into related quantities of force, pressure, power, energy, electrical current, voltage, resistance, etc., and thereby have a coherent system, with appropriate decimal multiples and submultiples.

In 1960 the Eleventh International General Conference on Weights and Measures carried out the first step refinement of the metric system to achieve uniformity in all countries and to satisfy the highest possible scientific requirements. This Eleventh International General Conference was followed by the Twelfth in 1964, the Thirteenth in 1967–68, and the Fourteenth in 1971.

The seven base units of the International Metric System (SI) are given in Table 3.1. The units of length, mass, and time have already been discussed and the current definitions for them have been given in Chapter 2.

The unit of temperature is the kelvin, defined as 1/273.16 part of the thermodynamic tempera-

TABLE 3.1

Base Units of the International Metric System (SI)

Quantity	Name of the Unit	Symbol
Length	meter	m
Mass	kilogram	kg
Time	second	s
Temperature	kelvin	K
Electric current	ampere	A
Luminous intensity	candela	cd
Amount of substance	mole	mol

ture of the triple point of water, and is discussed in detail in Chapter 4.

The unit of electric current is the ampere, defined as that constant current which, if maintained in two straight parallel conductors of infinite length, of negligible circular cross section, and placed 1 m apart in a vacuum, produces between these two conductors a force, due to their magnetic fields, equal to 2×10^{-7} newton per meter of length.

The unit of luminous intensity is the candela, defined as the luminous intensity, in the perpendicular direction, of a surface of 1/600,000 m^2 of a block body at the temperature of freezing platinum under a pressure of 101,325 N/m^2. This pressure is equivalent to 1 atm. The temperature of freezing platinum is 2,042 K.

The unit of amount of a substance is the mole, defined as that amount of the given substance that contains as many elementary entities (electrons, ions, atoms, molecules, or other specified particles) as there are atoms in 0.012 kg (12 g) of the pure nuclide Carbon-12.

The SI supplementary units for plane angle and solid angle are, respectively, radian, rad, and steradian, sr.

In Table 3.2 are given some derived units of the International Metric System, with their names and recommended symbols. No names have yet been proposed in the International Metric System for linear or angular velocity or acceleration, for area or volume or density, for kinematic or dynamic viscosity, for magnetic field strength, for luminance ($cd \cdot m^2$), for thermal conductivity, for wave number, etc.

Many of the base and derived units of the International Metric System (SI) have been in use in the cgs version of the metric system and are quite familiar. These units include the meter, kilogram, second, ampere, kelvin (formerly degree Kelvin), mole, joule, watt, coulomb, volt, ohm, and some others. However, the SI units of force and pressure are new and it is helpful to fix in one's mind the relationship between the cgs and SI forms of these properties. Following are these relations, including also the familiar unit of pressure, the atmosphere:

1 dyne (dyn) = 10^{-5} newton (N)
1 dyn · cm^{-2} = 10^{-1} N · m^{-2} = 10^{-1} pascal (Pa)
1 bar = 10^6 dyn · cm^{-2} = 10^5 N · m^{-2} = 10^5 Pa
1 atmosphere (atm) = 1,013,250 dyn · cm^{-2} = 1.013250 bar = 101325.0 N · m^{-2} = 101325.0 Pa

Table 3.3 gives the recommended decimal multiples and submultiples and the prefix names to be used.

TABLE 3.2

Some Derived Units of the International Metric System (SI)

Quantity	Name of the Unit	Symbol	Equivalence
Frequency	hertz	Hz	1 Hz = 1 s^{-1}
Force	newton	N	1 N = 1 $kg \cdot m \cdot s^{-2}$
Pressure	pascal	Pa	1 Pa = 1 $N \cdot m^{-2}$
Energy	joule	J	1 J = 1 $N \cdot m$
Power	watt	W	1 W = 1 $J \cdot s^{-1}$
Quantity of electricity	coulomb	C	1 C = 1 $A \cdot s$
Electrical potential or electromotive force	volt	V	1 V = 1 $W \cdot A^{-1}$
Electric resistance	ohm	Ω	1 Ω = 1 $V \cdot A^{-1}$
Electric conductance	siemens	S	1 S = 1 Ω^{-1}
Electric capacitance	farad	F	1 F = 1 $C \cdot V^{-1}$
Magnetic flux	weber	Wb	1 Wb = 1 $V \cdot s$
Magnetic flux density	tesla	T	1 T = 1 $Wb \cdot m^{-2}$
Inductance	henry	H	1 H = 1 $Wb \cdot A^{-1}$
Luminous flux	lumen	ℓm	1 ℓm = 1 cd·sr
Illumination	lux	ℓx	1 ℓx = 1 $\ell m \cdot m^{-2}$

TABLE 3.3

Recommended Decimal Multiples and Submultiples and the Corresponding Prefixes and Names

Factor	Prefix	Symbol	Meaning
10^{12}	tera	T	One trillion times
10^{9}	giga	G	One billion times
10^{6}	mega	M	One million times
10^{3}	kilo	k	One thousand times
10^{2}	hecto	h	One hundred times
10	deca	da	Ten times
10^{-1}	deci	d	One tenth of
10^{-2}	centi	c	One hundredth of
10^{-3}	milli	m	One thousandth of
10^{-6}	micro	μ	One millionth of
10^{-9}	nano	n	One billionth of
10^{-12}	pico	p	One trillionth of
10^{-15}	femto	f	One quadrillionth of
10^{-18}	atto	a	One quintillionth of

I. DISCUSSION

In the past century and a half, the enormous increases in the speeds of communication and transportation have brought all the countries of the world much closer together, with great increases in travel and trade among them and a resulting need for a common system of measurement. It was only natural then that the logical metric system, once having been developed, would find favor and adoption among nearly all the countries of the world.

It is expected that the International Metric System, a refinement of the old metric system, will more and more take over as the worldwide system of measurement in all aspects of human activity, in science, in engineering, in manufacturing, in industry, in commerce, in the marketplace, in the home, in sports, and in recreation.

For additional information on this subject, the reader is referred to the references, particularly De Simone[2] and Page and Vigoureux,[5] and to the headquarters of the organizations mentioned in the text. For some special information, the author is indebted to a friend and former associate at the National Bureau of Standards (U.S.), Louis E. Barbrow, coordinator of metric activities.

J. REFERENCES

1. Metric system; Metrology, in *Encyclopaedia Britannica,* Vol. 15, 14th ed., Encyclopaedia Britannica, New York, 1932.
2. **De Simone, D. V.,** *A Metric America,* National Bureau of Standards Special Publication 345, U.S. Government Printing Office, Washington, D.C., 1971.
3. **Helgren, F. J.,** *Metric Units of Measure,* 7th ed. Metric Association, Waukegan, Ill., 1967.
4. **Guillaume, C.-E. and Volet, C.,** in *International Critical Tables of Numerical Data, Physics, Chemistry, and Technology,* Vol. 1, Washburn, E. W., Ed., McGraw-Hill, New York, 1926, 1.
5. **Page, C. H. and Vigoureux, P.,** *The International System of Units (SI),* National Bureau of Standards Special Publication 330, U.S. Government Printing Office, Washington, D.C., 1972.
6. **Paul, M. A.,** The international system of units (SI), *J. Chem. Doc.,* 11, 3, 1971.

Chapter 4

THE BASIC SCALE OF TEMPERATURE

A. INTRODUCTORY COMMENTS

In this chapter we discuss the phenomenon of heat, discuss the concept of temperature and its quantitative measurement, give a brief history of the development of thermometers and of the scales of temperature associated with them, develop the "zero-pressure" gas scale of temperature and explain its equivalence to the thermodynamic scale of temperature, and review the scale of temperature in use internationally up to 1968, including the International Practical Temperature Scales of 1927 and 1948.

B. THE PHENOMENON OF HEAT

The recognition of the importance of thermal phenomena goes back to the Helenic period of Greeks, when it was noted that heat and cold were significant factors in natural occurrences. Later, distinction was made between the intensity of heat and its quantity, but the scientific elucidation of these properties remained at a standstill for lack of quantitative observations. It was not until the 16th Century that real understanding began to break through.

C. THE CONCEPT OF TEMPERATURE

Temperature is one of the most fundamental properties in the science of thermodynamics. The concept of temperature is easily comprehended through the human senses. Given a series of objects of the same material, each maintained at a different temperature, one can by the sense of feel tell whether one is hotter or colder than another. And, if two of them are maintained at the same temperature, one can distinguish this fact, within certain limits. Given five similar bodies, A, B, C, D, and E, maintained at temperatures T_A, T_B, T_C, T_D, and T_E, respectively, physical contact of the hand with each of these can permit one to classify them with respect to temperature. Suppose we find A to be warmer than B, B to be warmer than C, C to be neither warmer nor colder than D, and D to be neither warmer nor colder than E. From these observations, we may, within certain limits, deduce the following relations:

$$T_A > T_B > T_C; T_C = T_D = T_E; T_A > T_B > T_D; \quad T_A > T_B > T_E \tag{1}$$

It is important to note that these observations do not tell us anything about the quantity of heat, only about its intensity, the temperature.

Furthermore, we observe that when a hot body and a cold body are placed in contact with one another, heat energy flows from the hot body to the cold body, and the temperature of the former falls, and that of the latter rises, until equilibrium is reached with both bodies attaining substantially the same temperature.

D. THE QUANTITATIVE MEASUREMENT OF TEMPERATURE

Having a qualitative knowledge of temperature, one then desires to make some quantitative measurements. For this purpose, one may select a suitable substance having any readily measured property that changes monotonically with the temperature. The property measured must also be one that has reproducible values when the substance is repeatedly returned to the same temperature. For example, one may use for the thermometric property and the thermometric substance the following: the length of a given piece of copper rod; the volume of a given mass of mercury; the electrical resistance of a given piece of platinum wire; the volume, at constant pressure (or the pressure, at constant volume), of a given mass of oxygen gas. After the selection of the thermometric substance and the thermometric property, it is necessary to specify the mathematical function that is to relate the given property to the temperature. Such a function may be a simple linear one, in which the temperature is proportional to the length of the copper rod, the resistance of the platinum wire, or volume of mercury, or it may be a complicated mathematical function relating temperature and the given property.

Any two scales of temperature utilizing different thermometric substances, different thermometric properties, or both, will not be exactly the same. The reason is that different substances have different values of the percentage rate of change of a given property with temperature and because different properties of the same substance have different values of the percentage rate of change of the properties with temperature. Whenever the volume of a given fluid is measured in containers of different material, it is possible to have two slightly different scales of temperature even though the same property (volume) of the same substance is being used. For example, two scales of temperature, both utilizing the apparent expansion of mercury in glass, will be different if the glass containers in the two cases have different values of expansion with temperature.

It is obviously desirable to have a scale of temperature that is independent of a particular property of a particular substance.

E. EARLY THERMOMETERS AND THEIR SCALES OF TEMPERATURE

The record appears to indicate that the first real thermometer, an air themoscope, was made by Galileo in Italy about 1600. It is possible that certain other investigators in the period to 1622, namely Sanctorius, Fludd, and Drebbel, may also independently have invented the thermometer, but the record on this is not clear.

However, because of lack of agreement on a reproducible scale of temperature throughout the 17th Century, the thermometer did not become as useful an instrument as it should have.

Many of the early scales were divided into eight spaces. Sometimes these spaces were further subdivided into a few or many parts each. In 1615, Sagredo, friend of Galileo, divided the scale between the greatest heat of summer and the extreme cold of winter into 360 parts or "degrees." Other scales had various divisions: one of 50 parts, another of 100 or more parts, one of 7 major parts, another of 14 major parts, and two others of 12 major parts.

During the 17th Century, there was not only no uniformity with regard to the divisions on the scale, but also there was no general agreement as to the fixed points to be used in determining the limits of the scale. Among the fixed points proposed at that time were the following: winter heat and summer heat, the temperature of a deep cellar and that of melting butter or aniseed oil, and the freezing and boiling points of water. None received general approval.

In 1632 a French physician, Jean Rey, developed a thermometer in which liquid water was used as the thermometric substance, a rise of temperature resulting in expansion of the water from a reservior flask up into a long thin neck on the flask. The change from a gaseous to a liquid thermometric substance reduced, but did not wholly eliminate, the errors due to atmospheric pressure.

In 1641 Ferdinand II of Tuscany, Italy, one of the founders of the Florentine Accademia del Cimento, hermetically sealed his thermometers, using alcohol as the thermometric substance. He thus eliminated the effect of atmospheric pressure. This meant that greater accuracy was now attainable.

In 1665 Boyle, Hooke, and Huygens proposed that a single fixed point be used for calibrating thermometers, and that temperatures below and above this point be determined by appropriately proportioning the contraction or expansion of the thermometric substance.

Soon afterwards, a number of investigators suggested that two fixed points be used, with the interval between the two points to be divided in some manner to be agreed upon. In 1669 Fabri, member of the Florentine Accademia del Cimento, adopted the temperature of melting snow as the lower fixed point but retained the "greatest summer heat" as the higher fixed point. In 1688 Dalence assigned the value $-10°$ to the temperature of melting snow and $+10°$ to the temperature of melting butter. In 1694 Rinaldini, professor at Padua, Italy, proposed using the freezing and boiling temperatures of water as the two fixed points, with the interval divided into 12 parts.

A brief chronological record of subsequent events follows:

In 1701 Newton, physicist of England, proposed a scale with the freezing point of water as $0°$ and body temperature as $12°$.

In 1701 (approximately) Amontons, meteorologist of France, found that the thermal expansion of air is surprisingly uniform, and that, if a fixed mass of air is contained in a sealed vessel, its heat intensity (temperature) is related directly to the increase in pressure. He suggested a scale based on one fixed point, the boiling point of water, with the temperature to be measured as a proportionate decrease or increase in the pressure of the given mass of air at the given temperature. He found at the boiling point of water 73 units of pressure and at the freezing point of water 51 to 52 units of pressure, and inferred that there would be a zero temperature corresponding to zero pressure.

In 1702 Roemer, astronomer of Denmark, set the zero at the lowest temperature he could obtain with a freezing mixture of ice and salt and the upper fixed point as the boiling point of water.

In 1710 (approximately) Fahrenheit (Germany), maker of meteorological instruments, after a visit to Roemer, set his scale at 0° for the freezing mixture of ice and salt and the higher point as body temperature at 12°. Later, in 1717, he changed the latter point to 96°, so that his scale gave the freezing point of water as 32° and the boiling point of water as 212°.

In 1710 Elvius, of Sweden, is believed to have suggested assigning the values 0 and 100° to the freezing and boiling points of water.

In 1735 Reamur, of France, proposed a scale which gave 0° as the freezing point of water and 80° as the boiling point of water.

In 1740 Linnaeus, a Swedish botanist, also proposed using 0 and 100° as the values for the freezing and boiling points of water.

In 1742 Celsius, a Swedish astronomer, proposed a scale with the freezing point of water as 100° and the boiling point of water as 0°, being the first one, apparently, to subdivide the interval between these two fixed points into 100 units.

In 1742 (approximately) Martel, of Switzerland, was reported to have made a thermometer with the freezing point of water as 0°.

In 1743 Christian, of France, inverted the scale of Celsius, but did not take into account the effect of pressure on the boiling point of water.

In 1749 Stromer, of Sweden, inverted the scale of Celsius, taking the freezing point of water as 0° and the boiling point of water as 100°.

In 1779 Lambert, in 1787 Charles, in 1793 Volta, in 1801 Gay-Lussac, and in 1802 Dalton all promoted the use of the expansion of a gas as the basis for a thermometric system.

In 1848 Lord Kelvin (W. Thomson), of England, suggested using one realizable defining point, with the absolute zero of temperature as the origin. In the latter part of the 19th Century, Mendeleev, of Russia, is reported to have made the same suggestion.

In 1939, Giauque, of the U. S., revived the suggestion of Kelvin to use only one defining point, along with the absolute zero of temperature as the origin, to establish the absolute scale of temperature.

F. THE "ZERO-PRESSURE" GAS SCALE OF TEMPERATURE

It is found that scales of temperatures utilizing gases at low pressure (about 1 atm or less), involving either the change of volume at constant pressure or the change of pressure at constant volume, are very nearly (although not exactly) alike. The differences between such scales become less and less as the pressure is diminished. This behavior is corollary to the fact that all real gases approach a common limit in their relations of pressure, volume, and temperature as the pressure is diminished without limit. It is possible therefore to utilize any real gas as a standard thermometric substance by extrapolating its appropriate properties to zero pressure.

With a fixed, but not necessarily measured, quantity of any real gas at a given temperature, T, measurements of the pressure and volume are made at several finite low pressures. These values, in the form of the pressure-volume product, are extrapolated to zero pressure to obtain the value of the pressure-volume product at zero pressure at the given temperature for the quantity of gas involved. If these measurements are repeated at another temperature on the same quantity of the same gas, there is obtained a value of the pressure-volume product at zero pressure for the second temperature. A satisfactory and fundamental scale of temperature is established by letting these values of the pressure-volume product at zero pressure be proportional to the temperature on this zero-pressure gas scale, as:

$$(PV)_T^{P=0} = AT. \tag{2}$$

In Equation 2, A is the constant of proportionality for the quantity (preferably moles) of gas involved, and T is the temperature on this "zero-pressure" gas scale.

The next step is to evaluate the constant of proportionality. This may be done in either of two ways: (a) by defining the number of degrees between two selected fixed points that are realizable in the laboratory, or (b) by defining the absolute value of temperature to be assigned to one fixed point realizable in the laboratory with reference to the origin or zero on the zero-pressure gas scale of temperature. In the former method, which is the one used in the International Practical Temperature Scales up to 1960, the difference in the values of temperature between the two selected defining points is given and never changed, but the absolute values of the temperature of the two selected defining points may change as the result of improvements in experimentation with the zero-pressure gas thermometer. In the second method, which is the one that came into international acceptance in 1960, following the suggestions of Kelvin, Mendeleev, and Giauque,[10,11] the absolute value of the temperature of one defining point with reference to the origin or absolute zero is given and never changed, but the absolute value of the temperature of any other fixed point may change as the result of improvements in experimentation.

In the former method for defining the zero-pressure gas scale of temperature, the two realizable defining points which were employed to define the scale are (a) the "ice point," the temperature at which solid water is in thermodynamic equilibrium with liquid water in air at a pressure of 1 atm, and (b) the "steam point," the temperature at which liquid water is in thermodynamic equilibrium with gaseous water at a pressure of 1 atm. The difference in the temperature of these two points was defined as exactly 100 units or degrees.

If values of the pressure-volume product at zero pressure have been determined for the quantity of the given gas, at the ice and steam points, respectively, then for the former method of defining the zero-pressure gas scale of temperature the following relations hold:

$$T_{\text{steam}} - T_{\text{ice}} = 100 \text{ (defined constant)} \tag{3}$$

$$T_{\text{ice}} = 100 \frac{(PV)^{P=0}_{T_{\text{ice}}}}{(PV)^{P=0}_{T_{\text{steam}}} - (PV)^{P=0}_{T_{\text{ice}}}} \tag{4}$$

Equation 4 serves to evaluate the absolute temperature of the ice point on the zero-pressure gas scale of temperature, according to the former

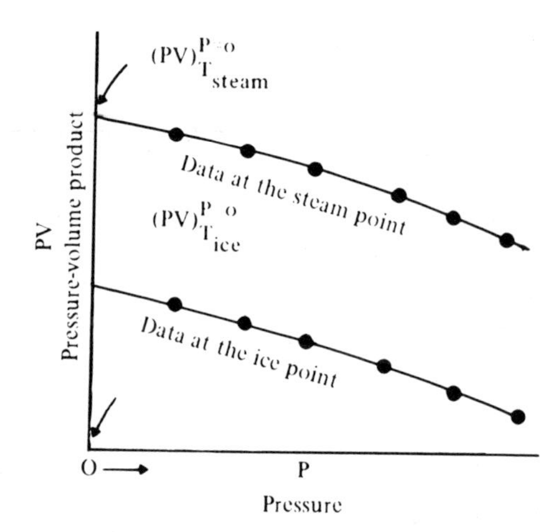

FIGURE 4.1. Simplified schematic diagram, showing the extrapolation of pressure-volume data to zero pressure. Not to scale. (From Rossini, F. D., *Chemical Thermodynamics,* John Wiley and Sons, New York, 1950. With permission.)

method of defining this scale of temperature by using two realizable defining points.

Figure 4.1 represents schematically how observations of the pressure-volume product for real gases are extrapolated to zero pressure. In principle, what is required is the measurement, on a fixed but not necessarily measured amount of a gas at a given temperature, of its pressure and volume at a series of pressures and the extrapolation of these values to zero pressure. Actually, the observations and calculations are rather involved, and the extrapolation to zero pressure is much more complicated than shown schematically in principle in Figure 4.1.

Figure 4.2 represents schematically the relation between the values of the temperature and the pressure-volume product at zero pressure for the zero-pressure gas scale of temperature defined by two realizable defining points.

Equation 4 may be easily derived from Figure 4.2 by making a proportion of the corresponding parts of the two similar right triangles:

$$\frac{\overline{AB}}{\overline{BC}} = \frac{\overline{AD}}{\overline{DE}} \tag{5}$$

or

$$T_{\text{ice}} = \overline{AB} = \overline{BC}\,\frac{\overline{AD}}{\overline{DE}}. \tag{6}$$

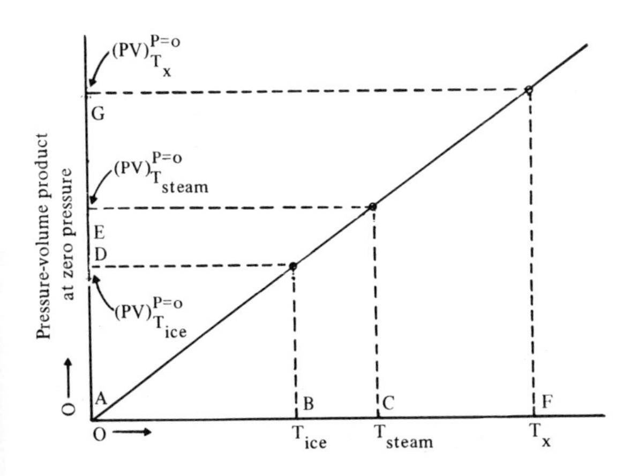

FIGURE 4.2. Simplified diagram, showing the relation between the pressure-volume product at zero pressure and the temperature on the zero-pressure gas scale with two defining points. Not to scale. (From Rossini, F. D., *Chemical Thermodynamics,* John Wiley and Sons, New York, 1950. With permission.)

Any other unknown temperature, including any additional fixed (but not defining) points, may be evaluated fundamentally from measurements on a given but not necessarily measured quantity of a suitable gas. Such measurements would determine the pressure-volume product at zero pressure at the given unknown temperature and at the two defining points, the ice and steam points. The value of the unknown temperature, T_x, on this scale would be given by the following relation:

$$T_x = 100 \frac{(PV)^{P=0}_{T_x}}{(PV)^{P=0}_{T_{steam}} - (PV)^{P=0}_{T_{ice}}} \,. \tag{5}$$

Equation 7 may be easily derived from Figure 4.2 by making a proportion of the corresponding parts of the two similar right triangles:

$$\frac{\overline{AF}}{\overline{BC}} = \frac{\overline{AG}}{\overline{DE}} \tag{8}$$

or

$$T_x = \overline{AF} = \overline{BC}\,\frac{\overline{AG}}{\overline{DE}} \,. \tag{9}$$

In such determinations, maximum precision and accuracy in the evaluation of the unknown temperature, T_x, are obtained by having the pressure-volume measurements at all three temperatures made with the same apparatus and procedure and by having the extrapolation of the experimental data to zero pressure made uniformly in the same way.

In the current method of defining the zero-pressure gas scale of temperature, using only one realizable defining point and the origin or absolute zero, the absolute temperature of the triple point of water is defined and any other unknown temperature, T_x, is evaluated fundamentally from this defined value of the triple point and the ratio of the pressure-volume product of a suitable gas at zero pressure at the unknown temperature and at the triple point. With the absolute temperature of the triple point of water taken as the best internationally accepted value, the relations involved are as follows:

$$T_{tp} = 273.16° \text{ (defined constant)} \tag{10}$$

$$T_x = 273.16 \frac{(PV)^{P=0}_{T_x}}{(PV)^{P=0}_{T_{tp}}} \,. \tag{11}$$

It may be noted here that the temperature at which solid water is in thermodynamic equilibrium with liquid water at saturation pressure, in the absence of air, which is the triple point, is 0.0100°C higher than the ice point involving water saturated with air at 1 atm. On this scale of temperature, the interval between the triple point and the steam point of water is 99.9900°.

On the basis of the present scale, defined with the triple point of water and the absolute zero of temperature, Figure 4.3 gives a pictorial representation of the principle involved. The distance AB is fixed by definition as 273.16 (exactly) units. The steam point then becomes simply a fixed point such as any other fixed point. Any "unknown" temperature, T_x, is evaluated by the relation:

$$T_x = \overline{AF} = \overline{AB}\,\frac{\overline{AG}}{\overline{AD}} \,. \tag{12}$$

One sees from the above that the system of using one realizable defining point, with the absolute zero of temperature as the origin, has great advantage over the system of using two realizable defining points. In the latter case the value of the absolute temperatures of the defining points will change with improved experimentations, whereas in the former case the absolute

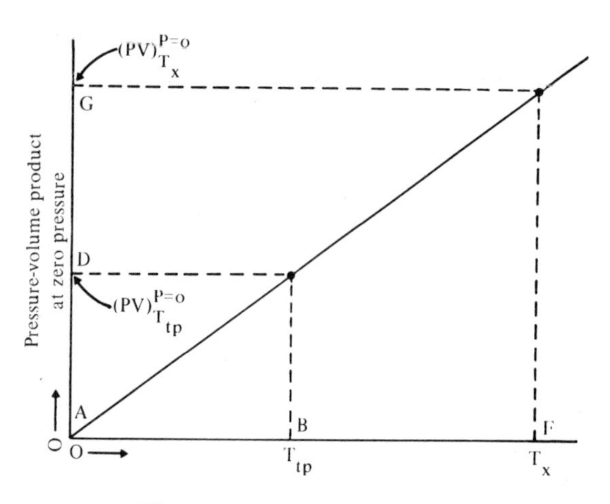

FIGURE 4.3. Simplified diagram, showing the relation between the pressure-volume product at zero pressure and the temperature on the zero-pressure gas scale with one defining point. Not to scale.

temperature of the defining point is fixed for all time.

G. THE PRACTICAL OR WORKING SCALE OF TEMPERATURE

The evaluation of temperatures by the use of a "zero-pressure" gas thermometer is essentially limited to the national standardizing laboratories and to certain other laboratories possessing the necessary apparatus and experience. The reason is that the use of a zero-pressure gas thermometric system with high precision and accuracy is laborious, difficult, and costly. For general use in science and technology, it has become necessary to establish a practical or working scale of temperature, accepted internationally.

To do this, one goes back to the selection of a specified thermometric substance, a specified property of that substance, and a specified mathematical function relating values of the given property of the given substance to values of the temperature.

Before making any measurements on the selected thermometric substance and property, it is necessary to determine experimentally the values of the temperature of a number of fixed points, in addition to the defining point, which are to be used in connection with the practical or working scale of temperature. The number of such fixed points will depend upon the number of constants to be evaluated in the selected mathematical function relating, with the necessary precision and accuracy, the values of the selected thermometric property of the selected thermometric substance to the values of the temperature of the selected fixed points in conjuction with the defining point.

To transfer the International Scale of Temperature to working laboratories, the "defining point" plus selected "fixed points" is thus needed. Reproducible fixed points are most easily realized by using the temperature of thermodynamic equilibrium of two or three phases of a given pure substance. The best fixed points are triple points, involving solid, liquid, and gaseous phases. The next best fixed points are freezing points, involving solid and liquid phases in air or other atmosphere at a given pressure. The more difficult fixed points to work with are boiling points, involving liquid and gaseous phases at a given pressure, because this temperature is much more sensitive to changes in pressure. Fixed points are selected on the basis of their reproducibility and the range of temperature to be covered.

H. THE INTERNATIONAL PRACTICAL TEMPERATURE SCALES OF 1948 AND 1927

The International Practical Temperature Scale of 1948 (IPTS-48) had the following values assigned for the absolute temperature of the two defining points, ice and steam, and the other fixed points, oxygen, sulfur, silver, and gold:

Substance	Equilibrium	Temperature (°C)
Oxygen	Liquid-gas, at 1 atm	-182.970
Water	Solid-liquid, in air at 1 atm	0.00 (exactly)
Water	Liquid-gas, at 1 atm	100.00 (exactly)
Sulfur	Liquid-gas, at 1 atm	444.600
Silver	Solid-liquid, at 1 atm	960.8
Gold	Solid-liquid, at 1 atm	1063.0

In 1960 the International Practical Temperature scale was modified by shifting to one defining point, the triple point of water. The value assigned to the absolute temperature of the triple point of water was 273.16 (exactly). The difference of temperature between the "ice point" and the "triple point" of water was known to be 0.0100 ± 0.0001°C.[12] This gives the absolute temperature of the "ice point" as 273.1500 ± 0.0001.

The next step was to select suitable thermometric substances, properties, and mathematical functions. It has been found that the electrical resistance of pure platinum wire increases in a roughly linear manner with temperature. With three or four constants in a second- or third-degree polynomial, it was possible to relate the electrical resistance of platinum to temperature on the above scale with considerable precision and accuracy, from -183 to 630°C, as follows:

For the range -183 to 0°C, the relation of resistance to temperature was

$$r_t = r_0(1 + at + bt^2 + ct^3) \tag{13}$$

where r_t and r_0 are the electrical resistances of the platinum resistance thermometer at the temperature t and at 0°C, respectively, and a, b, and c are constants evaluated from measurement of the resistance of the thermometer at the oxygen, steam, and sulfur points.

For the range 0 to 630°C, the relation of resistance to temperature was

$$r_t = r_0(1 + at + bt^2) \tag{14}$$

where the three constants are evaluated from measurement of the resistance of the thermometer at the ice, steam, and sulfur points.

For the range from 630 to 1,063°C, the electromotive force of a standard platinum vs. platinum-rhodium thermocouple was used for the working scale of temperature, with the following relation:

$$e = a + bt + ct^2. \tag{15}$$

Here e is the electromotive force of the standard thermocouple, one junction of which is kept at 0°C and the other at the given temperature t, and the three constants, a, b, and c, are evaluated from measurements of the electromotive force of the thermocouple at the antimony, silver, and gold points. The value of the temperature to be assigned to the antimony point in this calibration is evaluated with the platinum resistance thermometer for the given sample of antimony. (Pure antimony was assigned a freezing point of 630.5°C.)

Measurements of temperature below -183°C are made with gas thermometers, with resistance thermometers or thermocouples that have been calibrated against a gas thermometric system, or with resistance thermometers or thermocouples that have been calibrated against other resistance thermometers or thermocouples that have been calibrated against a gas thermometric system.

Measurements of temperature above the gold point are made with optical pyrometers. The measurements involve determination of the ratio of the intensity of monochromatic visible radiation of a given wave length emitted by a black body at the unknown temperature to the intensity of the same radiation of the same wave length emitted by a black body at the gold point. The unknown temperature T_X is evaluated from the radiation formula:

$$\frac{J_{T_X}}{J_{T_{Au}}} = \frac{e^{c_2/\lambda T_{Au}} - 1}{e^{c_2/\lambda T_X} - 1}. \tag{16}$$

In Equation 16, J_{T_X} and $J_{T_{Au}}$ are the radiant energies per unit wavelength interval at the given wave length, λ (in centimeters), emitted per unit time by unit area of a black body at the temperatures T_X and T_{Au}, respectively; T_X and T_{Au} are the absolute values of the unknown temperature and the temperature of the gold point, respectively; c_2 is the second radiation constant and is equal to hc/k, the value for which was taken as 1.438 cm · degrees for IPTS-48.

The International Practical Temperature Scale of 1927 (IPTS-27) had been changed as follows to produce IPTS-48. The temperatures assigned to the four fixed points were

Fixed point	1927 (°C)	1948 (°C)
Oxygen	-182.97	-182.970
Sulfur	444.60	444.600
Silver	960.5	960.8
Gold	1063.	1063.0

Also, the value of the second radiation constant,

c_2, was changed from 1.432 in IPTS-27 to 1.438 in IPTS-48.

I. THE THERMODYNAMIC SCALE OF TEMPERATURE

The "zero-pressure" gas scale of temperature described in this chapter is identical with the fundamental thermodynamic scale of temperature with the proportionality factor taken appropriately. This is shown as follows:

The ideal gas is defined by the two relations:

$$PV = RT \tag{17}$$

$$(\partial E/\partial V)_T = 0. \tag{18}$$

By the second law of thermodynamics, the absolute temperature was defined by the following relation for a reversible process:

$$\delta q = T dS. \tag{19}$$

By application of the first and second laws of thermodynamics to a substance participating in a reversible process involving only heat energy and PV work energy, the following relations can be derived:

$$\left[\frac{\partial E}{\partial V}\right]_T = -P + T\left[\frac{\partial P}{\partial T}\right]_V \tag{20}$$

$$\left[\frac{\partial H}{\partial P}\right]_T = V - T\left[\frac{\partial V}{\partial T}\right]_P . \tag{21}$$

Equations 20 and 21 may be rewritten explicitly in terms of pressure and volume as:

$$P = T\left[\frac{\partial P}{\partial T}\right]_V - \left[\frac{\partial E}{\partial V}\right]_T \tag{22}$$

and

$$V = T\left[\frac{\partial V}{\partial T}\right]_P + \left[\frac{\partial H}{\partial P}\right]_T . \tag{23}$$

It is important to note that Equations 22 and 23 have been derived independently of the equations defining the ideal gas, and the temperature is the thermodynamic temperature. If, therefore, Equations 22 and 23 are completely in accord with the equations defining the ideal gas, the two scales of temperature are proportional and may be considered identical if the constant of proportionality between the two scales is taken as unity. The accord of Equations 22 and 23 with the definition of the ideal gas may be shown by substituting the ideal gas relations into the right side of Equations 22 and 23 and seeing whether the resulting terms yield the pressure in Equation 22 and the volume in Equation 23. In Equation 23 the values of T, $(\partial P/\partial T)_V$, and $(\partial E/\partial V)_T$ from the relations of the ideal gas are introduced, the substitutions being T, R/V, and 0, respectively, which yield P for the sum of the two terms. Similarly, in Equation 23, the values of T, $(\partial V/\partial T)_P$, and $(\partial H/\partial P)_T$ are introduced from the relations for the ideal gas, the substitutions being T, R/P, and 0, which yield V for the sum of the two terms.

J. DISCUSSION

In this chapter we have discussed the phenomenon of heat and the concept of temperature and its quantitative measurement,[1-4] the early thermometers and their scales of temperature,[2-9] the "zero-pressure" gas scale of temperature,[1] the practical or working scale of temperature,[1] the International Practical Temperature Scales of 1948 and 1927,[12,13] and the thermodynamic scale of temperature.[1]

For additional information, the reader is directed to the references. Apart from the history, most of what is in this chapter is from a previous report of the author.[1]

K. REFERENCES

1. **Rossini, F. D.,** *Chemical Thermodynamics,* John Wiley and Sons, New York, 1950, chap. 3.
2. Heat, in *Encyclopaedia Britannica,* Vol. 11, 14th ed., Encyclopaedia Britannica, New York, 1932.
3. **Roller, D.,** *The Early Developments of Concepts of Temperature,* Harvard Case Histories in Experimental Science No. 3, Harvard University Press, Cambridge, Mass., 1950.
4. **Boyer, C. B.,** History of the measurement of heat. I. Thermometry and calorimetry, *Sci. Mon.,* 57, 442, 1943.
5. **Boyer, C. B.,** Early principles in the calibration of thermometers, *Am. J. Phys.,* 10, 176, 1942.
6. **Dorsey, N. E.,** Fahrenheit and Roemer, *J. Wash. Acad. Sci.,* 36, 361, 1946.
7. **Friend, J. N.,** The origin of Fahrenheit's scale, *Nature,* 139, 395, 1937.
8. **Meyer, K.,** Ole Romer and the thermometer, *Nature,* 82, 296, 1910.
9. **Meyer, K.,** Ole Romer's and Fahrenheit's thermometers, *Nature,* 139, 585, 1937.
10. **Thomson, W.,** *Phil. Mag.,* 33, 313, 1848.
11. **Giauque, W. F.,** *Nature,* 143, 623, 1939.
12. **Stimson, H. F.,** *J. Res. Nat. Bur. Stand.,* 42, 209, 1949.
13. **Burgess, G. K.,** *J. Res. Nat. Bur. Stand.,* 1, 635, 1928.

Chapter 5

THE INTERNATIONAL PRACTICAL TEMPERATURE SCALE OF 1968

A. INTRODUCTORY COMMENTS

In this chapter we describe in detail the current International Practical Temperature Scale of 1968 (IPTS-68). As given in the preceding chapter, this has superseded the International Practical Temperature Scale of 1948 as modified in 1960 (IPTS-48), which superseded the International Practical Temperature Scale of 1927 (IPTS-27). Included here are the essential features of IPTS-68, how this scale differs from IPTS-48, how the changes affect new measurements, and how measurements made on the basis of IPTS-48 can be corrected to the new basis of IPTS-68.

At its meeting in October, 1968, the International Committee on Weights and Measures approved the recommendations of its Consultative Committee on Thermometry and set up the International Practical Temperature Scale of 1968, to become effective January 1, 1969.

Additional details not given in this chapter may be found in the references given at the end of the chapter. Much of the material in this chapter is taken from Rossini.[1]

B. THE BASIS OF THE INTERNATIONAL PRACTICAL TEMPERATURE SCALE OF 1968

The International Practical Temperature Scale of 1968 has been set up in such a way that temperatures measured on it are very close to temperatures on the absolute thermodynamic scale, the difference between the two being within the limits of present-day measurements. IPTS-68 is constructed by assignment of an exact value of temperature to one defining point, the triple point of water, with reference to the absolute zero of temperature. Then selected best values of experimentally determined temperatures (as described in principle in the preceding chapter) are assigned to a number of reproducible primary fixed points. In contrast to IPTS-48 and IPTS-27, the new IPTS-68 also provides an array of secondary fixed points. These will be very useful to many investigators.

The basic elements, symbols, and nomenclature of IPTS-68 include the following:

1. The basic temperature is the absolute thermodynamic temperature, to which is given the symbol T.

2. The unit of temperature on this absolute thermodynamic scale is the kelvin, to which is given the symbol K.

3. The size of the kelvin, which is the unit degree on this absolute thermodynamic scale, is fixed by defining the temperature of the triple point of water as 273.16 (exactly) K.

4. A given numerical number of degrees above the absolute zero on this thermodynamic scale, say 298.16, is written as 298.16K, without the previously written superscript degree sign.

5. Temperatures on the Celsius scale (formerly called the Centigrade scale) are denoted by the symbol t.

6. The unit of temperature on the Celsius scale is the degree Celsius, to which is given the symbol °C.

7. The unit of temperature on the Celsius scale is exactly equal to the unit of temperature on the absolute thermodynamic scale:

$$\text{one degree Celsius} = \text{one kelvin (exactly)}. \quad (1)$$

8. Differences in temperature may be expressed either in kelvins or in degrees Celsius.

9. Temperatures on the Celsius* scale are related to temperatures on the absolute thermodynamic or Kelvin scale by the relation:

$$t = T - 273.15\text{K (exactly)} \quad (2)$$

The defining point and the primary and secondary fixed points involve thermodynamic equilibrium among three phases (gas-liquid-solid, triple point) or between two phases (liquid-solid, freezing point; gas-liquid, boiling point) of a pure substance. The defining point of IPTS-68 is a triple

*It may be noted here that the change from "Centigrade" to "Celsius" was agreed upon by the International Committee in order to avoid confusion with the meaning of "centigrade" in the French language, as 1/10,000 of a right angle.

point, while the primary and secondary fixed points involve triple points, freezing points, and boiling points. As mentioned in the preceding chapter, the reproducibility, precision, and accuracy of fixed points are normally best for triple points, less so for freezing points, and least for boiling points.

Interpolation between the defining point and adjacent primary fixed points, or between adjacent primary fixed points, is done by means of formulas relating readings with standard instruments and thermometers to values of temperature on IPTS-68.

C. THE DEFINING POINT: TRIPLE POINT OF WATER

The defining point for IPTS-68 is the triple point of water (equilibrium of water in three phases, solid, liquid and gas, in the absence of air or other substance). The value of temperature assigned to this point is 273.16K (exactly). This definition determines the size of the degree kelvin, as previously stated, and hence also the size of the degree Celsius.

The triple point of water replaced the freezing point of water (equilibrium between the solid and liquid phases of water, in the presence of air at a pressure of 1 atm) because the former is much more reproducible and stable than the latter.[2]

D. THE DIFFERENCE BETWEEN THE TRIPLE AND FREEZING POINTS OF WATER

In 1960, when the original International Practical Temperature Scale of 1948 was amended to produce what we now label as IPTS-48, and the triple point of water became the defining point, it was necessary to know rather well the difference between the triple point of water and the freezing point of water. Fortunately, this difference had been determined experimentally with considerable accuracy and precision.[2]

Following is a summary of the values assigned to the difference in temperature between the ice point and the triple point of water, together with their values on the Kelvin and Celsius scales:

By definition:

$$T_{tp} = 273.16 \text{ (exactly) K.} \quad (3)$$

By definition:

$$1 \text{ kelvin} = 1 \text{ degree Celsius (exactly).} \quad (1)$$

By experiment:

$$t_{tp} - t_{ice} = 0.0100 \pm 0.0001 \text{ °C.} \quad (4)$$

Hence:

$$T_{ice} = 273.1500 \pm 0.0001 \text{ K.} \quad (5)$$

By definition;

$$T_{tp} - t_{tp} = 273.15 \text{ (exactly).} \quad (6)$$

Hence, from Equations 3 and 6:

$$t_{tp} = 0.01 \text{ (exactly) °C.} \quad (7)$$

Hence, from Equations 4 and 7:

$$t_{ice}\ 0.0000 \pm 0.0001 \text{ °C.} \quad (8)$$

E. THE PRIMARY FIXED POINTS

The primary fixed points on IPTS-68, and the values assigned to them, are given in Table 5.1. The values actually selected are underlined – on the Kelvin scale below the freezing point of water and on the Celsius scale above the triple point of water. The defining point, the triple point of water, is set in bold type. The value for the freezing point of water (now a secondary fixed point) is included in this table simply for convenience. The values for the same temperature on the Kelvin and the Celsius scales differ by 273·15 (exactly).

In the last column of Table 5.1 are given the estimated uncertainties of the assigned values of temperature for the primary fixed points, referred to the thermodynamic scale of temperature.

As previously noted, all of the fixed points, as well as the defining point, involve thermodynamic equilibrium between two or three phases of a pure substance. In general, the triple point (equilibrium between solid, liquid, and gas phases) is the most reproducible and reliable, the freezing point (equilibrium between solid and liquid phase) is next, and the boiling point (equilibrium between liquid and gas phases) is next.

F. THE SECONDARY FIXED POINTS

In addition to the primary fixed points discussed above, the International Committee approved a large number of secondary reference points. The identification of these points, and the values of temperature assigned to them, are given in Table 5.2.

TABLE 5.1

Values of the Temperatures of the Primary Fixed Points on the International Practical Temperature Scale of 1968, and Their Estimated Uncertainties in Terms of Thermodynamic Temperatures[a]

Substance	Equilibrium	T^g K	t^g °C	Estimated uncertainty K
Hydrogen[b]	Solid-liquid-gas	13.810	−259.340	±0.010
Hydrogen[b]	Liquid-gas, at $\frac{25}{76}$ atm	17.042	−256.108	±0.010
Hydrogen[b]	Liquid-gas, at 1 atm	20.280	−252.870	±0.010
Neon[c]	Liquid-gas, at 1 atm	27.102	−246.048	±0.010
Oxygen	Solid-liquid-gas	54.361	−218.789	±0.010
Oxygen	Liquid-gas, at 1 atm	90.188	−182.962	±0.010
(Water)[d]	(Solid-liquid, in air at 1 atm)	(273.1500)	(0.0000)	(±0.0001)
Water[e]	Solid-liquid-gas	**273.16**	**0.01**	**exact**
Water	Liquid-gas, at 1 atm	373.150	100.000	±0.005
(Tin)[f]	Solid-liquid, at 1 atm	505.118	231.968	±0.015
Zinc	Solid-liquid, at 1 atm	692.73	419.58	±0.03
Silver	Solid-liquid, at 1 atm	1,235.08	961.93	±0.20
Gold	Solid-liquid, at 1 atm	1,337.58	1,064.43	±0.20

[a]In interpreting the facts given in this table the following points are to be noted. The abbreviation "atm" means the standard atmosphere defined as 1013250 dyn cm^{-2} or 101325 N m^{-2}. The numbers 25/76 and 1 before atm are atmosphere defined as 1013250 dyn cm^{-2} or 101325 N m^{-2}. The numbers 25/76 and 1 before atm are taken as exact. Near 1 atm the freezing point of various metals changes in amounts ranging only 0·00001 to 0·0001 degree per 0·01 atm change in pressure. The depth of immersion of the thermometer in the liquid phase of various metals affects the temperature only by amounts ranging from 0·00001 to 0·0001 degree per 1-cm change in depth of immersion.

[b]The hydrogen referred to in this table means "equilibrium" hydrogen, which, at any given temperature, is in equilibrium with respect to the ortho and para forms of hydrogen. At the normal boiling point, 1 atm, the composition of "equilibrium" hydrogen is 0·21% ortho and 99·79% para, while at room temperature it is near 75% ortho and 25% para. The latter mixture, remained unchanged in composition, has a normal boiling point, at 1 atm, which is 0·12 degree above that of "equilibrium" hydrogen. Equilibrium between the ortho and para forms of hydrogen is achieved by use of ferric hydroxide as a catalyst.

[c]Neon, which is largely ^{20}Ne, normally contains 0·0026 mol fraction of ^{21}Ne and 0·088 mol fraction of ^{22}Ne.

[d]The value for the temperature of the ice point is a secondary reference point, but is included here for the convenience of the reader (see Table 5.2).

[e]The water should have the isotopic composition of ocean water. The extreme difference, in temperature of the triple point of water from natural sources, ocean water, and continental surface water has been found to be about 0·00025 degree.

[f]The freezing point of tin may be used in place of the normal boiling point of water as one of the primary fixed points.

[g]The defining point, the triple point of water, is in boldface. The values for the same temperature on the two scales differ by 273·15 (exactly). The number of significant figures given here varies in a few cases from the official report.[1]

From Rossini, F. D., *Pure Appl. Chem.*, 22, 557, 1972. With permission.

TABLE 5.2

Values of the Temperatures of the Secondary Reference Points on the International Practical Temperature Scale of 1968[a]

Substance	Equilibrium	T K	t °C
Hydrogen, "normal"[b]	Solid-liquid-gas	13.956	-259.194
Hydrogen, "normal"[b]	Liquid-gas, 1 atm	20.397	-252.753
Neon	Solid-liquid-gas	24.555	-248.595
Nitrogen	Solid-liquid-gas	63.148	-210.002
Nitrogen	Liquid-gas, 1 atm	77.348	-195.802
Carbon dioxide	Solid-gas, 1 atm	194.674	- 78.476
Mercury	Solid-liquid, 1 atm	234.288	- 38.862
Water	Solid-liquid, in air, 1 atm	273.1500	0.0000
Phenoxybenzene (diphenylether)	Solid-liquid-gas	300.02	26.87
Benzoic acid	Solid-liquid-gas	395.52	122.37
Indium	Solid-liquid, 1 atm	429.784	156.634
Bismuth	Solid-liquid, 1 atm	544.592	271.442
Cadmium	Solid-liquid, 1 atm	594.258	321.108
Lead	Solid-liquid, 1 atm	600.652	327.502
Mercury	Liquid-gas, 1 atm	629.81	356.66
Sulfur	Liquid-gas, 1 atm	717.824	444.674
Cu-Al, eutectic	Solid-liquid, 1 atm	821.38	548.23
Antimony	Solid-liquid, 1 atm	903.89	630.74
Aluminum	Solid-liquid, 1 atm	933.52	660.37
Copper	Solid-liquid, 1 atm	1,357.6	1,084.5
Nickel	Solid-liquid, 1 atm	1,728	1,455
Cobalt	Solid-liquid, 1 atm	1,767	1,494
Palladium	Solid-liquid, 1 atm	1,827	1,554
Platinum	Solid-liquid, 1 atm	2,045	1,772
Rhodium	Solid-liquid, 1 atm	2,236	1,963
Iridium	Solid-liquid, 1 atm	2,720	2,447
Tungsten	Solid-liquid, 1 atm	3,660	3,387

[a]See the footnotes to Table 5.1.
[b]"Normal" hydrogen is hydrogen having a composition of ortho and para hydrogen corresponding to that of "equilibrium" hydrogen at room temperature (see footnote b of Table 5.1).

From Rossini, F. D., *Pure Appl. Chem.*, 22, 557, 1972. With permission.

G. THERMOMETRIC SYSTEMS

Having established the necessary array of fixed points, the next step is to specify the thermometric systems, that is, the thermometric substances to be used and the properties to be measured, along with the standard measuring instruments, for the several ranges into which the total scale is subdivided. IPTS-68 is based on the use of only three thermometric systems:

1. From 14 to 904K, the platinum resistance thermometer, with measurement of the electrical resistance of a coil of pure, strain-free, annealed platinum.
2. From 904 to 1338K, the thermocouple, of platinum and an alloy of 90% platinum and 10% rhodium, with measurement of the electromotive force.
3. Above 1338K, the optical pyrometer, using the Planck radiation formula, with measurement of the intensity of radiation.

Table 5.3 summarizes the specifications for the several ranges of IPTS-68, showing the temperature covered in each range, the calibrating points for the given range, and the measuring instrument for that range.

H. REALIZATION OF THE SCALE OVER THE RANGE 13.810 to 903.89K

For this range, where the platinum resistance thermometer applies, the basic measurement is the resistance ratio W. For the unknown temperature T_x, the resistance ratio is

$$W_{T_x} = R_{T_x}/R_{273.15}. \quad (9)$$

TABLE 5.3

Specifications for the Several Ranges of the International Practical Temperature Scale of 1968

Range of temperature K	Range of temperature °C	Calibrating points	Measuring instrument
13.810–20.280	-259.340–-252.870	Triple pt, H_2 Boiling pt, H_2, 25/76 atm Boiling pt, H_2, 1 atm Triple pt, H_2O	Platinum resistance thermometer
20.280–54.361	-252.870–-218.789	Boiling pt, H_2, 1 atm Boiling pt, Ne, 1 atm Triple pt, O_2 Triple pt, H_2O	Platinum resistance thermometer
54.361–90.188	-218.789–-182.962	Triple pt, O_2 Boiling pt, O_2, 1 atm Triple pt, H_2O	Platinum resistance thermometer
90.188–273.15	-182.962–0.00	Boiling pt, O_2, 1 atm Triple pt, H_2O Boiling pt, H_2O, 1 atm	Platinum resistance thermometer
273.15–903.89	0.00–630.74	Triple pt, H_2O Boiling pt, H_2O, 1 atm (or freezing pt, Sn) Freezing pt, Zn	Platinum resistance thermometer
903.89–1,337.58	630.74–1,064.43	(Freezing pt, Sb)[a] Freezing pt, Ag Freezing pt, Au	Thermocouple: platinum and 10% Rh–90% Pt
1,337.58 and above	1,064.43 and above	Freezing pt, Au, with Planck radiation equation	Optical pyrometer

[a]See text following Equation 22.

From Rossini, F. D., *Pure Appl. Chem.*, 22, 557, 1972. With permission.

Here, R_{Tx} is the resistance at T_x and $R_{273.15}$ is the resistance at 273.15 (exactly) K or 0.00 (exactly) °C, which is 0.01 (exactly) kelvin below the triple point of water. This is also the freezing point of water (within 0.0001°C). One condition applies to W, namely, that its value at 373.15K (100.00°C) must not be less than 1.39250.

Below 273.15K, the relation between temperature and resistance of the thermometer is obtained from a reference function and certain specified deviation equations. For the range 13.81 to 273.15K, a reference function, *W-CCT*, has been tabulated as a function of temperature to provide interpolation with a precision of 0.0001K.[4] With *W-CCT* thus defined, T_x is evaluated from the relation:

$$W_{T_x} = R_{T_x}/R_{273.15} = (W\text{-}CCT)_{T_x} + \Delta W_{T_x}. \qquad (10)$$

Here, ΔW_{Tx} is determined separately for each of four subranges, as follows.

From 13.810 to 20.280K,

$$\Delta W_{T_x} = A_1 + B_1 T_x + C_1 T_x^2 + D_1 T_x^3 \qquad (11)$$

where the constants, A_1, B_1, C_1, and D_1, are evaluated from observations at the four calibrating points specified for this subrange (see Table 5.3): by the measured deviations at the triple point of "equilibrium" hydrogen (13.810K), the temperature of 17.042K (the boiling point of "equilibrium" hydrogen at 25/76 atm), and the normal boiling point of "equilibrium" hydrogen* (20.280K), and by the temperature derivative of ΔW_{Tx} at the normal boiling point of "equilibrium" hydrogen (20.280K) as derived from the following Equation 12.

From 20.280 to 54.361K,

*As used here, the terminology 'normal boiling point' means the boiling point (thermodynamic equilibrium between the liquid and gas phases) at a pressure of exactly one atmosphere.

$$\Delta W_{T_x} = A_2 + B_2 T_x + C_2 T_x^2 + D_2 T_x^3 \quad (12)$$

where the constants, A_2, B_2, C_2, and D_2, are evaluated from observations at the four calibrating points specified for this subrange (see Table 5.3): by the measured deviations at the normal boiling point of "equilibrium" hydrogen (20.280K), the normal boiling point of neon (27.102K), and the triple point of oxygen (54.361K), and by the temperature derivative of ΔW_{Tx} at the triple point of oxygen (54.361K) as derived from the following Equation 13.

From 54.361 to 90.188K,

$$\Delta W_{T_x} = A_3 + B_3 T_x + C_3 T_x^2 \quad (13)$$

where the constants A_3, B_3, and C_3, are evaluated from observations at the three calibrating points specified for this range (see Table 5.3): by the measured deviations at the triple point of oxygen (54.361K) and the normal boiling point of oxygen (90.188K), and by the temperature derivative of ΔW_{Tx} at the normal boiling point of oxygen (90.188K) as derived from the following Equation 14.

From 90.188 to 273.15K,

$$\Delta W_{T_x} = A_4 T_x + C_4 T_x^3 (T_x - 100) \quad (14)$$

where the constants, A_4 and C_4, are evaluated from observations at the two calibrating points specified for this range (see Table 5.3): by the measured deviations at the normal boiling point of oxygen (90.188K) and the normal boiling point of water (373.150K or 100.000°C).

For the range 273.15 to 903.89K, the following equations are used:

$$T_x = t_x + 273.15 \text{ (exactly) K} \quad (15)$$

$$t_x = t' + 0.045 \left(\frac{t'}{100}\right)\left(\frac{t'}{100} - 1\right)\left(\frac{t'}{419.58} - 1\right)\left(\frac{t'}{630.74} - 1\right) \text{°C} \quad (16)$$

$$t' = \frac{1}{\alpha}(W_{t'} - 1) + \delta\left(\frac{t'}{100}\right)\left(\frac{t'}{100} - 1\right). \quad (17)$$

In Equation 17, the constants, α and δ, are evaluated from measurements of the resistance ratio, $W_{t'}$, at the normal boiling point of water (100.000°C or 373.150K) or the freezing point of tin (231.968°C or 505.118K) and the freezing point of zinc (419.58°C or 692.73K). Here we have, for the normal boiling point of water,

$$W_{373.15} = R_{373.15}/R_{273.15} \quad (18)$$

for the freezing point of tin,

$$W_{505.118} = R_{505.118}/R_{273.15} \quad (19)$$

and for the freezing point of zinc,

$$W_{692.73} = R_{692.73}/R_{273.15}. \quad (20)$$

I. REALIZATION OF THE SCALE OVER THE RANGE 903.89 to 1377.58K

For the range 903 89 to 1337 58K, the following equations are used:

$$T_x = t_x + 273.15 \text{ (exactly)} \quad (21)$$

$$E_{t_x} = a + bt_x + ct_x^2. \quad (22)$$

Here, E_{tx} is the electromotive force of the standard thermocouple of platinum and an alloy of 90% platinum-10% rhodium, with one junction at 273.15K, or zero degrees C, and the other at the unknown temperature t_x. The constants, *a, b,* and *c*, are evaluated from the measured values of the electromotive force at the freezing point of gold (1337.58K or 1064.43°C), the freezing point of silver (1235.08K or 961.93°C), and at 630.74 ± 0.20°C (903.89K) as determined by the platinum resistance thermometer as specified in the foregoing. (The temperature, 903.89K or 630.74°C, is a secondary reference point, corresponding to the freezing point of antimony.)

The specifications require that the standard thermocouple shall be annealed and that the purity of the platinum wire shall be such that the resistance ratio $W_{373.15}$, equal to $R_{373.15}/R_{273.15}$, shall be not less than 1.3920. The companion wire shall be an alloy containing 90% platinum and 10% rhodium, by weight.

Further, the thermocouple shall satisfy the following relations:

At the gold point (1337.58K or 1064.43°C),

$$E_{Au} = 10300 \pm 50 \text{ microvolts}. \quad (23)$$

For the difference in electromotive force between the gold point and the silver point (1235.08K or 961.933°C),

$$E_{Au} - E_{Ag} = 1183 + 0.158\,(E_{Au} - 10300) \pm 4 \text{ microvolts}. \quad (24)$$

For the difference in electromotive force between the gold point and 630.74°C (or 903.89K),

$$E_{Au} - E_{903.89} = 4766 + 0.631\,(E_{Au} - 10300) \pm 8 \text{ microvolts}. \quad (25)$$

J. REALIZATION OF THE SCALE ABOVE 1337.58K

Above the gold point (1337.58K), the unknown temperature, T_x, is defined by the Planck radiation formula:

$$J_{T_x}/J_{T_{Au}} = [\exp(c_2/\lambda T_{Au}) - 1]/[\exp(c_2/\lambda T_x) - 1] \quad (26)$$

In this equation: T_x and T_{Au} refer to the unknown temperature and the temperature of the gold point on the Kelvin scale, respectively; J is the spectral concentration, $\partial L \partial \lambda$ of the radiant energy, L, per unit wavelength interval at the given wavelength, λ, emitted per unit time by unit area of a black body at the given temperature; c_2 is the second radiation constant with the following value:

$$c_2 = 0.014388 \text{ meter kelvin}. \quad (27)$$

The measurements involve determination, with an optical pyrometer, of the ratio of the intensity of monochromatic visible radiation of a given wavelength emitted by a black body at the unknown temperature to the intensity of the same radiation of the same wavelength emitted by a black body at the gold point.

K. RECOMMENDATIONS REGARDING APPARATUS, METHODS, AND PROCEDURES*

The official publication[3] on the International Practical Temperature Scale of 1968 gives some detailed recommendations on apparatus, methods, and procedures, covering the following items:

- Standard resistance thermometer
- Standard thermocouple
- Triple point and the normal boiling point of "equilibrium" hydrogen
- Normal boiling point of neon
- Triple point and normal boiling point of oxygen
- Normal boiling point of water
- Freezing point of tin
- Freezing point of zinc
- Freezing point of silver
- Freezing point of gold.

L. NUMERICAL DIFFERENCES BETWEEN THE INTERNATIONAL PRACTICAL TEMPERATURE SCALE OF 1968 AND THAT OF 1948

Values of the numerical differences between IPTS-68 and IPTS-48, $T_{68} - T_{48}$, for the range 90 to 10,000K, are given in Tables 5.4 and 5.5.** These values are taken from the paper of Douglas,[6] which values are equivalent, to the same number of significant figures, with the values of $T_{68} - T_{48}$ given in the official report[3] of the International Committee, for the range (up to 4,000°C) covered in the latter report.

Douglas[6] also provides values of $d(T_{68} - T_{48})/dT$, the change of $T_{68} - T_{48}$ with temperature up to 10,0000K. His values are given in Tables 5.6 and 5.7.

M. THE PROBLEM OF CONVERTING EXISTING CALORIMETRICALLY DETERMINED DATA TO THE BASIS OF THE INTERNATIONAL PRACTICAL TEMPERATURE SCALE OF 1968

With new calorimetric data being determined under the International Practical Temperature Scale of 1968, it becomes necessary to arrange for the conversion of existing calorimetrically determined data, obtained under IPTS-48, to the basis of IPTS-68.

Douglas[6] has prepared a report which gives detailed and exact formulas for making such conversions for calorimetric data on enthalpy, heat capacity, and entropy. Douglas[6] also gives equations for converting extrapolated data, on the basis of the "T^3," or the other theoretical or empirical relation, from a lowest temperature of measurement to zero K, for enthalpy, heat capacity, and entropy.

The problem is to convert experimentally measured calorimetric data, obtained at a given numerical value of temperature on IPTS-48, to the same numerical value of temperature on IPTS-68. Letting the given numerical value of temperature on IPTS-48 be T'_{48} and the same numerical value of temperature on IPTS-68 be T''_{68}, one can therefore write

$$T''_{68} = T'_{48}. \quad (28)$$

There is one point on IPTS-48 which has not only the same numerical value but also exactly the same temperature as on IPTS-68. This point is the triple point of water, at 273.16K. For convenience, the enthalpy, H, the heat capacity, C_p, and the entropy, S, at the same numerical value of temperature on IPTS-68 and IPTS-48, are designated as H'' and H', C''_p and C'_p, and S'' and S', respectively.

Also at a given actual temperature, the value on IPTS-68 will be T_{68} and that on IPTS-48 will be T_{48}.

Conversion of calorimetric data on enthalpy – Douglas[6] gives an exact equation, of an infinite series

*The material in Sections 5.E through 5.K is taken from Rossini, F. D., *Pure Appl. Chem.*, 22, 557, 1972, and is reprinted with permission.

**For detailed information on the several scales of temperature in the range 14 to 90K in use before 1968, and their relation to IPTS-68, see Reference 5.

TABLE 5.4

Differences in the Values of Temperature, over the Range 90K to the Gold Point, 1337.58K, Given by IPTS-68 and IPTS-48, Reported as $T_{68} - T_{48} = \Delta$*

Units: T_{68} in degrees on the Kelvin scale of IPTS-68; Δ in millikelvins

T_{68}	Δ	T_{68}	Δ	T_{68}	Δ	T_{68}	Δ	T_{68}	Δ
90	+ 8	140	− 9	280	− 3	480	45	860	134
92	11	145	− 5	290	− 7	490	49	880	160
94	13	150	0	300	− 9	500	53	900	194
96	13	155	+ 5	310	−10	520	60	920	245
98	12	160	10	320	−10	540	66	940	300
100	11	165	15	330	−10	560	70	960	354
102	9	170	20	340	− 9	580	74	980	409
104	6	175	24	350	− 7	600	76	1,000	464
106	4	180	27	360	− 4	620	77	1,020	519
108	1	185	30	370	− 1	640	77	1,040	575
110	− 1	190	32	380	+ 2	660	76	1,060	631
112	− 4	195	33	390	6	680	76	1,080	687
114	− 6	200	34	400	10	700	75	1,100	743
116	− 8	210	33	410	15	720	74	1,150	886
118	−10	220	30	420	19	740	75	1,200	1,029
120	−11	230	25	430	24	760	77	1,250	1,173
124	−13	240	20	440	28	780	81	1,300	1,319
128	−14	250	14	450	32	800	88	1,337.58	1,430
132	−13	260	7	460	37	820	98		
136	−11	270	2	470	41	840	113		

*The values in this table are rounded from the tabulation of Douglas.[6]

From Rossini, F. D., *Pure Appl. Chem.*, 22, 557, 1972. With permission.

TABLE 5.5

Differences in the Values of Temperature, from the Gold Point, 1337.58K, to 10,000K, Given by IPTS-68 and IPTS-48, Reported as $T_{68} - T_{48} = \Delta$*

Units: T_{68} in degrees on the Kelvin scale of IPTS-68; Δ in kelvins

T_{68}	Δ	T_{68}	Δ	T_{68}	Δ	T_{68}	Δ	T_{68}	Δ
1,337.58	1.43	1,850	2.34	2,800	4.6	3,900	8.0	5,000	12.3
1,350	1.45	1,900	2.44	2,900	4.8	4,000	8.3	5,500	14
1,400	1.53	1,950	2.54	3,000	5.1	4,100	8.7	6,000	17
1,450	1.61	2,000	2.65	3,100	5.4	4,200	9.1	6,500	19
1,500	1.70	2,100	2.86	3,200	5.7	4,300	9.4	7,000	22
1,550	1.78	2,200	3.08	3,300	6.0	4,400	9.8	7,500	25
1,600	1.87	2,300	3.31	3,400	6.3	4,500	10.2	8,000	27
1,650	1.96	2,400	3.55	3,500	6.6	4,600	10.6	8,500	30
1,700	2.05	2,500	3.79	3,600	7.0	4,700	11.0	9,000	33
1,750	2.15	2,600	4.0	3,700	7.3	4,800	11.4	9,500	37
1,800	2.24	2,700	4.3	3,800	7.6	4,900	11.8	10,000	40

*The values in this table are from the tabulation of Douglas.[6]

From Rossini, F. D., *Pure Appl. Chem.*, 22, 557, 1972. With permission.

TABLE 5.6

Differences in the Values of the Temperature Derivatives Over the Range 90K to the Gold Point, 1337.58K, given by IPTS-68 and IPTS-48, Reported as $d(T_{68} - T_{48})/dT = d\Delta/dT$[d]

Units: T_{68} in degrees on the Kelvin scale of IPTS-68; $d\Delta/dT$ in millikelvins per Kelvin

T_{68}	$d\Delta/dT$	T_{68}	$d\Delta/dT$	T_{68}	$d\Delta/dT$	T_{68}	$d\Delta/dT$	T_{68}	$d\Delta/dT$
90	2.2	140	0.71	280	−0.41	480	0.41	860	1.2
92	1.3	145	0.89	290	−0.29	490	0.39	880	1.5
94	0.5	150	1.00	300	−0.18	500	0.37	900	1.9
96	−0.1	155	1.03	310	−0.08	520	0.3	903.89	2.0[a] 2.7[b]
98	−0.6	160	1.02	320	+0.01	540	0.3	920	2.7
100	−0.89	165	0.96	330	0.09	560	0.2	940	2.7
102	−1.12	170	0.87	340	0.16	580	0.1	960	2.7
104	−1.24	175	0.75	350	0.23	600	0.1	980	2.7
106	−1.30	180	0.61	360	0.28	620	0.0	1,000	2.8
108	−1.30	185	0.47	370	0.33	640	0.0	1,020	2.8
110	−1.25	190	0.32	380	0.36	660	0.0	1,040	2.8
112	−1.16	195	0.17	390	0.39	680	0.0	1,060	2.8
114	−1.06	200	0.05	400	0.42	700	0.0	1,080	2.8
116	−0.92	210	−0.20	410	0.44	720	0.0	1,100	2.8
118	−0.78	220	−0.38	420	0.45	740	+0.1	1,150	2.9
120	−0.63	230	−0.51	430	0.45	760	0.1	1,200	2.9
124	−0.30	240	−0.59	440	0.45	780	0.3	1,250	2.9
128	0.00	250	−0.62	450	0.45	800	0.4	1,300	2.9
132	+0.28	260	−0.61	460	0.44	820	0.6	1,337.58	3.0[b] 1.6[c]
136	0.52	270	−0.54	470	0.43	840	0.9		

[a]From the range of platinum thermometer.
[b]From the range of the thermocouple (Pt and 10%Rh-90%Pt).
[c]From the range of the radiation scale.
*The values in this table are from the tabulation of Douglas.[6]

From Rossini, F. D., *Pure Appl. Chem.*, 22, 557, 1972. With permission.

TABLE 5.7

Differences in the Values of the Temperature Derivatives, from the Gold Point, 1337.58K, to 10,000K, Reported as $d(T_{68} - T_{48})/dT = d\Delta/dT$*

Units: T_{68} in degrees on the Kelvin scale of IPTS-68; $d\Delta/dT$ in millikelvins per Kelvin

T_{68}	$d\Delta/dT$	T_{68}	$d\Delta/dT$	T_{68}	$d\Delta/dT$	T_{68}	$d\Delta/dT$	T_{68}	$d\Delta/dT$
1,337.58	3.0[b]	1,700	1.9	2,200	2.2	3,500	3	7,000	5
	1.6[c]	1,800	1.9	2,300	2.3	4,000	4	8,000	6
1,400	1.6	1,900	2.0	2,400	2.4	4,500	4	9,000	6
1,500	1.7	2,000	2.1	2,500	2.5	5,000	4	10,000	6
1,600	1.8	2,100	2.2	3,000	3	6,000	5		

[b,c]See corresponding footnotes to Table 5.6.
*The values in this table are from the tabulation of Douglas.[6]

From Rossini, F. D., *Pure Appl. Chem.*, 22, 557, 1972, With permission.

type, for calculating the conversion of calorimetric data on enthalpy from a given numerical value of temperature, T'_{48}, on IPTS-48, to the same numerical value of temperature T''_{68}, on IPTS-68.

However, it is shown that all of the terms of this equation beyond the first are normally negligible in actual practice. Consequently, one obtains for the correction in enthalpy:[6]

$$\delta H = H'' - H' = - C_p(T_{68} - T_{48}). \quad (29)$$

Here C_p is the measured value of the heat capacity at the given temperature, $T_{68} - T_{48}$ is taken from Tables 5.4 and 5.5, and $H'' - H'$ is the correction to be added to the value of enthalpy previously reported under IPTS-48.

Conversion of calorimetric data on heat capacity – Douglas[6] gives an exact equation, also of an infinite series type, for calculating the conversion of calorimetric data on heat capacity from a given numerical value of temperature, T'_{48}, on IPTS-48, to the same numerical value of temperature, T''_{68}, on IPTS-68.

However, it is shown that the following approximate equation, adequate for nearly all cases encountered in actual practice, can be derived:[6]

$$\delta C_p = C''_p - C'_p = - C_p \mathrm{d}(T_{68} - T_{48})/\mathrm{d}T - (T_{68} - T_{48})\mathrm{d}C_p/\mathrm{d}T. \quad (30)$$

Here C_p and $T_{68} - T_{48}$ have the same significance as above for enthalpy: values of $\mathrm{d}(T_{68} - T_{48})/\mathrm{d}T$, the rate of change of $T_{68} - T_{48}$ with temperature, are given in Tables 5.6 and 5.7, $\mathrm{d}C_p/\mathrm{d}T$ is the rate of change of C_p with temperature, and $C''_p - C'_p$ is the correction to be added to the value of heat capacity previously reported under IPTS-48.

Conversion of calorimetric data on entropy – Douglas[6] gives an exact equation, also of an infinite series type, for calculating the conversion of calorimetric data on entropy from a given numerical value of temperature, T'_{48}, on IPTS-48, to the same numerical value of temperature, T''_{68}, on IPTS-68.

However, it is shown that the following approximate equation, adequate for nearly all cases in actual practice, can be derived:[6]

$$\delta S = S'' - S' = - \int_0^T [(T_{68} - T_{48}) C_p/T^2]\, \mathrm{d}T - (T_{68} - T_{48}) C_p/T. \quad (31)$$

Here C_p and $T_{68} - T_{48}$ have the same significance as above for enthalpy, the integration is taken from 0 to T, and $S'' - S'$ is the correction to be added to the value of entropy previously reported under IPTS-48.

N. CONVERSION OF *P-V-T* DATA TO THE BASIS OF THE INTERNATIONAL PRACTICAL TEMPERATURE SCALE OF 1968

Angus[7] has prepared a report in which he discusses the correction of experimental *P-V-T* data obtained under IPTS-48 to the basis of IPTS-68. In general, the procedure involves the conversion of data labeled for a given numerical value of temperature, T'_{48}, under IPTS-48, to the same numerical value of temperature, T''_{68}, under IPTS-68. In setting up the procedure for correcting the existing experimental data on *P-V-T* measurements to the new IPTS-68, one simply traces the effect of shifting the temperature by the amount $T_{68} - T_{48}$ (from Tables 5.4 and 5.5) at each temperature of measurement, utilizing values of $\mathrm{d}(T_{68} - T_{48})/\mathrm{d}T$ (from Tables 5.6 and 5.7) as appropriate. Since this report is mainly concerned with calorimetric data, the reader is referred to the report of Angus[7] for further details.

O. THERMODYNAMIC PROPERTIES CALCULATED STATISTICALLY*

In the case of values of thermodynamic properties calculated statistically from spectroscopic and other molecular data, with the proper values of the fundamental physical constants being used, no changes are necessary as a result of the introduction of the new IPTS-68. Following is the reasoning: (a) the values of thermodynamic properties calculated statistically are specified for temperatures on the thermodynamic scale, and (b) the values of temperature on the new IPTS-68 are as near the corresponding temperatures on the thermodynamic scale as is possible at the present time.

P. DISCUSSION

International agreement on the use of one defining point to establish the practical scale of temperature has led to considerable improvement in the recording and tabulating of absolute temperatures on the Kelvin scale and the conversion of such temperatures to the Celsius scale. From 1920 to 1960, the absolute temperature of the ice point of water was recorded, by responsible investigators and bodies, at various times and in various countries, at values ranging from 273.09 to 273.20. The variations resulted in annoying differences in the comparison of data obtained in various laboratories of the world. Hopefully, now, there will be much less confusion in this regard.

It is interesting to note the changes in the

*The material in Sections 5.M through 5.O is taken from Rossini, F. D., *Pure Appl. Chem.*, 22, 557, 1972, and is reprinted with permission.

values of the principal fixed points assigned in IPTS-27, IPTS-48, and IPTS-68. The oxygen point changed from −182.97 to −182.970 to −182.962 °C, the sulfur point from 444.60 to 444.600 °C to abandonment, the silver point from 960.5 to 960.8 to 961.93 °C, and the gold point from 1,063 to 1,063.0 to 1,064.43 °C. At the same time, the second radiation constant, c_2, changed from 1.432 to 1.438 to 1.4388 cm·degrees.

According to Gray and Finch,[8] the overall uncertainty in the calibration of platinum thermometers, with a direct calibration from the National Bureau of Standards, is as follows: 10 to 15K, ± 0.05 to ± 0.06K; 15 to 90K, ± 0.01 to ± 0.02K; oxygen point, ± 0.005K; triple point of water, ± 0.0003K; freezing point of tin, ± 0.002K; freezing point of zinc, ± 0.002K. For thermocouples[8], the NBS uncertainties are as follows: 1,100 °C, ± 0.6K; 1,250 °C, ± 0.8K; 1,800 °C, ± 2.0K; 2,800 °C, ± 3.8K; 3,500 °C, ± 8K. Transfer of such calibrations to working thermometers will increase the practical uncertainty for each thermometer, sometimes quite significantly.[8] For commercial thermocouples, the uncertainty may range from ± 1.6K at 630 °C up to 3.6K at 1,450 °C; for commercial automatic optical pyrometers, the uncertainty may range from ± 2K at 1,100 °C to ± 25K at 3,500 °C, while for commercial visual optical pyrometers, the uncertainty may range from ± 5K at 1,100 °C to ± 50K at 3,500 °C.

Utilizing the information given in this chapter, the practicing scientist or engineer can determine readily what effect the shift from IPTS-48 to IPTS-68 has on his current experimental measurements involving temperature, what he must do to have his measurements of temperature conform to IPTS-68, and how he should proceed to correct his previous data, obtained under IPTS-48, to the basis of IPTS-68.

On the basis of past experience, it appears that no significant changes in the International Practical Temperature Scale of 1968 will likely be made before about 1985.

Detailed information regarding the calibration of thermometers of various levels of accuracy and precision for the several ranges may be obtained from the national standardizing laboratory in each country, such as the National Bureau of Standards in the United States, the National Physical Laboratory in the United Kingdom, and similar bodies in France, Germany, the U.S.S.R., Japan, etc.

The author is indebted to R. P. Hudson, T. B. Douglas,[6] C. W. Beckett, and S. Angus for providing manuscripts in advance of publication of the report by Rossini,[1] and to G. Waddington, G. T. Furukawa, J. L. Riddle, and E. F. Westrum, Jr. for reviewing that report.[1] Additional information may be obtained from Reference 9.

Q. REFERENCES

1. **Rossini, F. D.,** *Pure Appl. Chem.*, 22, 557, 1972; *J. Chem. Thermodynamics*, 2, 447, 1972.
2. **Stimson, H. F.,** *J. Wash. Acad. Sci.*, 35, 201, 1945.
3. *The International Practical Temperature Scale of 1968,* Comptes Rendus de la 13 éme Conference General des Poids et Mesures, 1967–68, Annex 2; *Metrologia,* 5, 33, 1969.
4. **Bedford, R. E., Preston-Thomas H., Durieux, M., and Muijlwijk, R.,** *Metrologia,* 5, 45, 1969.
5. **Bedford, R. E., Durieux, M., Muijlwijk, R., and Barber, C. R.,** *Metrologia,* 5, 47, 1969.
6. **Douglas, T. B.,** *J. Res. Nat. Bur. Stand.*, 73A, 451, 1969.
7. **Angus, S. A.,** Report PC/D26, available from S. Angus, Imperial College of Science and Technology, London, Engl.
8. **Gray, W. T. and Finch, D. I.,** *Phys. Today,* 24, 32, 1971.
9. *Temperature – Its Measurement and Control in Science and Industry,* Vol. 4, Instrument Society of America, Pittsburgh, 1971.

Chapter 6

THE SCALE OF PRESSURE

A. INTRODUCTORY COMMENTS

Temperature and pressure are two of the most important variables that affect the properties of substances. In some ways, one can visualize, in the classical sense, the state of infinite pressure as a state of zero motion and ordered arrangement, somewhat analogous to the state of absolute zero of temperature.[4] We have already discussed the scale of temperature. In this chapter we discuss the scale of pressure, including the matter of units, pressures below atmospheric pressure, pressures above atmospheric pressure, the primary pressure scale, basic pressure-measuring instruments, fixed points on the scale of pressure and their detection, summary of the fixed points, and the galvanic cell as a device for measuring pressure.

Much of the information given in this chapter is derived from McNish,[1] Decker, Bassett, Merrill, Hall, and Barnett,[2] and Heydemann.[3]

B. UNITS OF PRESSURE

Pressure is force per unit area. In the "cgs" system of units, the unit of force is the dyne, defined as the force required to impart an acceleration of 1 cm/sec to a mass of 1 g, and the unit of pressure is 1 dyn/cm^2.

In the International Metric System (SI), the unit of force is the newton, N, where (see Chapter 3, Section H)

$$1 \text{ N} = 10^5 \text{ dyn} \tag{1}$$

and the unit of pressure is the pascal, Pa, where

$$1 \text{ Pa} = 1 \text{ N m}^{-2}. \tag{2}$$

Since

$$1 \text{ bar} = 10^6 \text{ dyn cm}^{-2} \tag{3}$$

it follows that

$$1 \text{ bar} = 10^5 \text{ N m}^{-2} = 10^5 \text{ Pa}. \tag{4}$$

The atmosphere is defined as

$$1 \text{ atm} = 1{,}013{,}250 \text{ dyn cm}^{-2}. \tag{5}$$

Hence,

$$1 \text{ atm} = 1.013250 \text{ bar} = 101325.0 \text{ Pa} \tag{6}$$

and

$$1 \text{ bar} = 0.986923 \text{ atm}. \tag{7}$$

Other conversion factors are given in Chapter 12.

For pressures below atmospheric pressure, most investigators use "mm Hg," or preferably "torr," as the unit of pressure. Under standard conditions, the relation is

$$1 \text{ atm} = 760 \text{ mm Hg} = 760 \text{ torr}. \tag{8}$$

C. BASIC MEASUREMENTS OF PRESSURE BELOW ATMOSPHERIC PRESSURE

Basic measurements of pressure below atmospheric pressure are made with a series of instruments, each of which, with some overlap, covers a range of pressure. From McNish,[1] amended by recent information from Heydemann,[3] the several instruments and their estimated accuracy are summarized as follows:

1. The air piston gage is used in the range from several atmospheres down to about 30 torr, with an estimated accuracy of 20 ppm above 1 atm down to 200 ppm at 30 torr.

2. The standard mercury barometer is used from 1 atm (± 40 ppm) down to near 0.01 torr where the uncertainty is of the order of the pressure itself. At 1 torr, the uncertainty is about 1%.

3. The micrometer-mercury manometer is used from about 40 torr (± 200 ppm) down to about 0.005 torr, where the uncertainty is of the order of the pressure itself. At 1 torr, the uncertainty is about 0.3%.

4. The interferometer-mercury manometer, under development, can be used from the range of one to several atmospheres, where the

uncertainty is about 1 ppm, down to about 0.005 torr, where the uncertainty is of the order of the pressure itself. At 1 torr, the uncertainty is about 0.2%.

5. The micrometer-oil manometer is used in the range from about 0.03 torr (± 0.4%) down to about 0.0001 torr, where the uncertainty is of the order of the pressure itself. At 0.01 torr, the uncertainty is about 1%.

6. The interferometer-oil manometer, under development, can be used from about 0.05 torr (± 0.01%) down to about 10^{-5} torr, where the uncertainty is of the order of the pressure itself. At 0.005 torr, the uncertainty is about ± 0.1%.

7. Potential devices for measuring very low pressures include the following: the diaphragm gage, in the range from 10^{-3} to 10^{-5} torr; the quartz vane gage, in the range from 10^{-6} to 10^{-8} torr; and the electrostatic vane gage, in the range from 10^{-7} to 10^{-9} torr. The uncertainties in such devices range from about 1% at the higher pressure to 100% at the lower pressure.

8. Conventionally, the ranges of pressure may be identified as follows:[1] low vacuum, 1 atm to 1 torr; medium vacuum, 1 to 10^{-3} torr; high vacuum, 10^{-3} to 10^{-6} torr; very high vacuum, 10^{-6} to 10^{-9} torr; ultra high vacuum, 10^{-9} torr and lower.

D. DEVICES AVAILABLE FOR BASIC MEASUREMENTS OF PRESSURE ABOVE ATMOSPHERIC PRESSURE

The devices available for basic measurements of pressure above atmospheric pressure include the following:[2,3]

1. The mercury manometer is used up to pressures of 10 bar (or more), with an uncertainty of ± 0.0003%.

2. The free-piston or dead-weight gage is used for pressures in the range 10 to about 26,000 bars, with an uncertainty of ±0.002% at the lower pressures up to ± 0.1% at the higher pressures.

3. The piston-cylinder gage (with packing) is used for pressures above 26,000 bars, with an uncertainty of ± 1 to ± 2%.

E. THE MERCURY MANOMETER

Because of its inherent simplicity, and the fact that the relevant characteristics leading to knowledge of the force and the area in a given apparatus can be measured accurately with high precision, the mercury manometer is the most suitable basic instrument for measurements of pressure over the range for which it is applicable. The fundamental properties involved are the height of the column of mercury, the density of the mercury, and the gravitational field at the given point. For greatest precision, the density of the mercury, which is a function of both temperature and pressure, must be known over the entire column, from base to top, with the desired accuracy.

In the period 1840 to 1900, a number of mercury manometers were constructed, with one end open to the atmosphere, with the density of the mercury measured at one level and assumed constant throughout the column, and with heights up to 300 m.

In modern practice, columns of several meters in height are preferred, using mercury of the highest purity, with well-controlled temperature baths containing the manometers, with precise and accurate equipment for measuring differences in height of the mercury meniscus in the two legs, and with all connections carefully monitored to assure calculation of all pressures.

Multiple-tube manometers and differential manometers were developed early.[2] In 1894, Stratton[5] suggested the use of a multiple-tube manometer, consisting of a number of alternating columns of mercury and of a liquid of lower density, as water or an organic liquid substance. In 1915, Holborn and Schultze[6] described a differential mercury manometer consisting of a single mercury column to each end of which is attached a free-piston gage. In 1927, Keyes and Dewey[7] constructed a differential manometer capable of measuring pressures up to 600 bars with a reported accuracy of ± 0.0001%. In 1931, Meyers and Jessup[8] built a five-column multiple manometer for use up to 15 bars, with an accuracy better than ± 0.0001%; operation of these five columns as a unit in a differential thermometer permitted extension of the range up to 75 bars with an accuracy of about ± 0.003%. In 1954, Roebuck and Obser[9] measured pressures up to 200 bars, with an accuracy better than ± 0.0001%, using a multiple nine-column manometer, each column 17 m in length, with the temperature controlled to 0.3°C. In 1954, Bett, Hayes, and Newitt[10] (see also, in 1963, Bett and Newitt[11]) constructed a differential manometer 9 m high, with temperature controlled to about 0.02°C, with high purity mercury and accurate measurements of the density of the mercury, for measurements up to 2,500 bars.

The manometers used at the higher pressures indicated in the several foregoing investigations were very useful and helpful in the basic calibration of free-piston gages to lead into the next higher range of pressure.

F. THE FREE-PISTON GAGE

With the free-piston gage, the force and the area are directly measurable, and hence this device serves also as a scale of pressure over the range for which it is applicable, provided the effective area can be determined accurately. Continuous rotation of the piston improves the performance of the free-piston gage. At the lower pressures of its range, the free-piston gage can be compared directly against the basic mercury manometer. Such intercomparisons of free-piston gages against mercury manometers have been reported by several investigators:[2] Michels,[12] Keyes and Dewey,[7] Beattie and Bridgeman,[13] Roebuck and Obser,[9] Bett and Newitt,[11] Cross,[14] Johnson, Cross, Hill, and Bowman,[15] Johnson and Newhall,[16] Dadson,[17] and Konyaev.[18]

In 1909–11, Bridgman[47] designed a reentrant cylinder system in which the pressure exerted prevented enlargement of the space between the rotating piston and its cylindrical housing, thus eliminating leakage around the piston.

The work of Johnson and Heydemann[19] indicates that a primary scale of pressure based upon the free-piston gage, appropriately calibrated at lower pressures against the mercury manometer, can be extended to 26,000 bars.

G. THE PISTON-CYLINDER GAGE

The piston-cylinder gage is used for pressures above 26,000 bars. This gage, which has either piston packing or a solid-medium pressure environment, is presently the best approximation to a primary scale of pressures above 26,000 bars.[2] Although some attempts have been made to use such a scale at pressures below 26,000 bar,[20,21] it appears that the future usefulness of this instrument as a scale of pressure will be limited to pressures above 26,000 bars.[2]

H. DESIRED CHARACTERISTICS OF FIXED POINTS ON THE SCALE OF PRESSURE

In order to transfer the fundamental scale of pressure to working laboratories, it is desirable to have a set of fixed points which serve to transfer the scale of pressure from the standardizing laboratory to the working laboratory.

As in the case of the fixed points for the International Practical Temperature Scale referred to in the preceding chapter, such fixed points are best based on solid phase transitions of adequately pure substances at given constant temperatures. Such phase changes can be liquid-solid or solid-solid, preferably representing thermodynamic equilibrium of two given phases of a pure substance at a given temperature and pressure.

Such fixed points must be ones in which equilibrium is set up with reasonable speed, so that overly long periods of time are not required. This implies reasonably rapid nucleation of the given solid species.

In addition to the foregoing requirements, the given phase change must be susceptible to detection by significant change in some property that can be measured without too much difficulty and with adequate precision.

Investigations on fixed points and their characteristics have been conducted by a large number of investigators:[2] Corll and Warren,[22] Corll,[23] Davidson and Lee,[24] Heydemann,[25] Zeto, Vanfleet, Hryckowian, and Bosco,[26] Gschneider, Elliott, and McDonald,[27] Brandt and Ginzburg,[28] Bundy,[29] Darnell,[30] Bridgman,[31] and Dadson and Greig.[32]

I. DETECTION OF PHASE CHANGES IDENTIFYING FIXED POINTS

In order to have practical usefulness, the given substances must have a property the value of which changes quite significantly at the pressure at which the phase stable at the lower pressure begins to change into the phase stable at the higher pressure, with the temperature being kept constant. Also, the property in question must be one that can be readily measured with adequate precision.

Among the properties used to detect such phase changes are the following:[2]

1. Changes in electrical resistance have been investigated by Bridgman[33] and Balchan and Drickamer.[34]

2. Changes in volume have been studied by Bridgman.[35]

3. Changes in optical properties have been studied by several investigators: Balchan[36] and Weir, Van Valkenburg, and Lippincott[37] on refractive index; Klyuev[38] on absorption; and Bassett, Takahashi, and Stook[39] on reflectance.

4. Changes in crystal structure have been investigated by Jeffery, Barnett, Vanfleet, and Hall[40] and by Jamieson[41] with X-rays; by Brugger, Bennion, and Worlton[42] and by Bennion, Worlton, Peterson, and Brugger[43] with neutron diffraction.

5. Measurements of differential thermal analysis were made by Kennedy and La Mori.[44]

6. Changes in magnetic properties were studied by Cleron, Coston, and Drickamer.[45]

7. Changes in ultrasonic velocities were investigated by Hagelberg, Holton, and Kao.[46]

Of the foregoing properties, measurements of the changes in electrical resistance and in volume have been the most accurate, precise, and useful, although several of the others are proving to be practical.

It is to be noted that indication of a phase change by some of the foregoing properties does not always correspond to thermodynamic equilibrium. In systems subjected to high pressure, where high strains are present, local regions may exist in the material where the pressure is sufficient to nucleate a critical volume and initiate the transition before the average pressure over the entire sample has reached the equilibrium value. The result is a transition that is not sharp but is broad with respect to pressure.

The purity of the material is important; an impurity of less than about 0.1% appears to be negligible in comparison with other possible errors. The constancy of temperature required for a given substance depends upon the value of dP/dT, the change of pressure with temperature. This is related to the enthalpy of transition, ΔH, and the increase in volume arising from the transition, ΔV, in accordance with the following equation:

$$dP/dT = \Delta H/T\Delta V . \qquad (9)$$

J. SUMMARY OF THE FIXED POINTS OF THE SCALE OF PRESSURE

The data summarized in Table 6.1 (from Reference 2) give a good picture of the accord achieved by various investigators in determining the pressure assignable to the transition of mercury from the liquid phase to the alpha solid phase at 0°C. One sees that, excluding the value

TABLE 6.1

Summary of the Data Obtained at 0°C for the Pressure of the Transition, Hg (liq) = Hg (solid, α)

Investigation	Pressure (bars)	Uncertainty (bars)	Method of detection
Bridgman[47]	7,492	± 72	Volume
Johnson and Newhall[16]	7,568	± 50	Volume
Zhokhovskii[48]	7,565.8	± 3.0	Pressure drop at initiation of transition
Zhokhovskii et al.[49]	7,569.7	± 2.3	Pressure drop at initiation of transition
Newhall, Abbot, and Dunn[50]	7,566.2	± 3.4	Volume
Dadson and Greig[32]	7,569.2	± 1.2	Electrical resistance
Yasunami[51]	7,571.0	± 1.2	Latent heat
Cross[52]	7,567.4	± 1.6	Electrical resistance
Weighted average: (excluding Bridgman)	7,569.2	± 1.5	

From Decker, D. L., Bassett, W. A., Merrill, L., Hall, H. T., and Barnett, J. D., *J. Phys. Chem. Ref. Data,* 1, 773, 1972. With permission.

TABLE 6.2

Summary of Fixed Points on the Scale of Pressure

Substance	Transition	Pressure (kbars)	Temperature (°C)
Mercury	Liquid = solid, α	7.5692 ± 0.0015	0
Bismuth	Solid I = solid II	25.50 ± 0.06	25
Thallium	Solid II = solid III	36.7 ± 0.3	25
Cesium	Solid II = solid III	42.5 ± 1.0	25
Cesium	Solid III = solid IV	43.0 ± 1.0	25
Barium	Solid I = solid II	55 ± 2	25
Bismuth	Solid III = solid V	77 ± 3	25
Tin	Solid I = solid II	100 ± 6	25
Iron	Solid α = solid ϵ	126 ± 6	25
Barium	Solid II = solid III	140 ± ?	25
Lead	Solid I = solid II	120–160	25

From Decker, D. L., Bassett, W. A., Merrill, L., Hall, H. T., and Barnett, J. D., *J. Phys. Chem. Ref. Data,* 1, 773, 1972. With permission.

reported by Bridgman in 1911, the data of the other seven investigations, reported in the years 1953 to 1968, have a weighted average of 7,569.2 ± 1.5 bar at 0°C, which is an uncertainty of only ± 0.02%.

For fixed points at higher pressures, the uncertainty increases significantly with pressure.

Table 6.2 summarizes the fixed points on the scale of pressure that currently have some standing in the community of scientific investigators working at high pressures.[2] In connection with the values given in Table 6.2, the following should be noted. For the Bismuth (I=II) transition, Heydemann[25] found the value 25.499 ± 0.060 kbars. The report of Decker et al.[2] gives the foregoing, and the following as selected "best" values: Cesium (II=III), 41.2 ± 1.2 kbars; Cesium (III-IV), 42.0 ± 1.2 kbars; Barium (I=II), 55.3 ± 1.2 kbars; Bismuth (III=V), 76.7 ± 1.8 kbars.

K. THE GALVANIC CELL AS A DEVICE FOR MEASURING PRESSURE

For a galvanic cell, the basic thermodynamic relations are the following.[53]

At constant temperature, for a given substance,

$$dG = V dP. \tag{10}$$

and for a given chemical reaction, such as may occur in a galvanic cell,

$$d(\Delta G) = \Delta V dP. \tag{11}$$

Here G is the Gibbs free energy, V is the molal volume, P is the pressure, and the symbol Δ means "increase in the value of."

For a galvanic cell with electromotive force, $\mathcal{E}$, it can be shown that, at constant pressure and temperature,

$$\Delta G = -n\mathcal{F}\mathcal{E}. \tag{12}$$

where n is the number of moles of electrons per mole of reaction and $\mathcal{F}$ is the Faraday Constant.

Combination of Equations 11 and 12 yields:

$$-n\mathcal{F}d\mathcal{E} = (\Delta V)dP. \tag{13}$$

Solving Equation 13 for dP, one obtains:

$$dP = -n \quad (1/\Delta V)d\mathcal{E} . \tag{14}$$

Integrating from zero pressure to a given pressure P, one obtains:

$$P = -n\mathcal{F}\int_0^P (1/\Delta V)d\mathcal{E} . \tag{15}$$

Equation 15 says that, for measurements at constant temperature, a plot of $1/\Delta V$ on the scale of ordinates against $\mathcal{E}$ on the scale of abscissas, for the given values of pressure, gives a curve the area under which, from $\mathcal{E}$ at pressure zero to $\mathcal{E}$ at pressure P, gives the value of the pressure.

A specific example of a galvanic cell with solid electrolytes is the following:

$$Pb/PbCl_2 \parallel AgCl/Ag . \tag{16}$$

In this cell, the half reaction at the left side is

$$Pb + 2Cl^- = PbCl_2 + 2e^- \quad (17)$$

the half reaction at the right side is

$$2e^- + 2AgCl = 2Ag + 2Cl^- \quad (18)$$

and the complete reaction is

$$Pb + 2AgCl = PbCl_2 + 2Ag \quad (19)$$

with all substances in the solid state.

It is seen from the foregoing that the development of a galvanic cell with solid electrolytes to serve as a device for measuring high pressures is an attractive possibility.[2] What is required is to have a galvanic cell with solid electrolytes, with a known chemical reaction, that is reversible and reproducible. In Equation 15, the value of $\mathcal{E}$ is measured electrically, and the value of ΔV is calculated from separate measurements of the molal volume (density), as a function of pressure, of each of the reactants and products of the known reaction.

L. DISCUSSION

Decker et al.[2] report the following: (a) improved techniques in the use and calibration of the free-piston gage have improved the reliability of measurements of pressure in the range up to 10 kbars; (b) the accuracies currently attainable are of the order of ± 0.02% at 8 kbars, ± 0.25% at 25 kbars, ± 2% at 50 kbars, and ± 4% at 100 kbars.

It is clear that continued investigation will reduce the uncertainties now associated with the measurement of pressures in the high ranges. Further, it would be helpful to the entire scientific and technical community in the communication of experimental data at high pressures if international accord could be reached, through the appropriate bodies, to produce an International Practical Pressure Scale in the same manner as the International Practical Temperature Scale described in the preceding chapter.

The author is indebted to friends at the National Bureau of Standards for special information on this subject: Charles W. Beckett, thermodynamics expert; Peter L. M. Heydemann, chief of the section on pressure.

M. REFERENCES

1. **McNish, A. G.,** Fundamentals of measurement, *Electron Technology, Science and Engineering Series,* 53, 113, 1963.
2. **Decker, D. L., Bassett, W. A., Merrill, L., Hall, H. T., and Barnett, J. D.,** *J. Phys. Chem. Ref. Data,* 1, 773, 1972.
3. **Heydemann, P. L. M.,** National Bureau of Standards, Washington, D.C., private communication, 1973.
4. **Lewis, G. N.,** *Z. Phys. Chem.,* 130, 532, 1927.
5. **Stratton, S. W.,** *Phil. Mag.,* 38, 160, 1894.
6. **Holborn, L. and Schultze, H.,** *Ann. Phys.* (Leipzig), 47, 1089, 1915.
7. **Keyes, F. G. and Dewey, J.,** *J. Opt. Soc. Am.,* 14, 491, 1927.
8. **Meyers, C. H. and Jessup, R. S.,** *J. Res. Nat. Bur. Stand.,* 6, 1061, 1931.
9. **Roebuck, J. R. and Obser, H. W.,** *Rev. Sci. Instrum.,* 25, 46, 1954.
10. **Bett, K. E., Hayes, P. F., and Newitt, D. M.,** *Phil Trans. R. Soc. Lond.,* 247, 59, 1954.
11. **Bett, K. E. and Newitt, D. M.,** *The Physics and Chemistry of High Pressures,* Society of Chemical Industry, London, 1963, 99.
12. **Michels, A.,** *Ann. Phys.* (Leipzig), 72, 285, 1923; 73, 577, 1924.
13. **Beattie, J. A. and Bridgeman, O. C.,** *J. Am. Chem. Soc.,* 49, 1665, 1927.
14. **Cross, J. L.,** *Nat. Bur. Stand. Monogr.,* 65, 1, 1964.
15. **Johnson, D. P., Cross, J. L., Hill, J. D., and Bowman, H. A.,** *Ind. Eng. Chem.,* 49, 2046, 1957.
16. **Johnson, D. P. and Newhall, D. H.,** *Trans. Am. Soc. Mech. Eng.,* 75, 301, 1953.
17. **Dadson, R. S.,** *Nature,* 176, 188, 1955; *Proc. Joint Conf. Thermodynamic and Transport Properties of Fluids,* Institution of Mechanical Engineers, Lond., 1958, 37.
18. **Konyaev, V. S.,** *Instrum. Exp. Tech. U.S.S.R.,* 1961, 728, 1961.
19. **Johnson, D. P. and Heydemann, P. L. M.,** *Rev. Sci. Instrum.,* 38, 1294, 1967.
20. **Kennedy, G. C. and Lamoir, P. N.,** *J. Geophys. Res.,* 67, 851, 1962.

21. **Boyd, F. R. and England, J. L.,** *J. Geophys. Res.,* 65, 741, 1960.
22. **Corll, J. A. and Warren, W. E.,** *J. Appl. Phys.,* 36, 3655, 1965.
23. **Corll, J. A.,** *J. Appl. Phys.,* 38, 2708, 1967.
24. **Davidson, T. E. and Lee, A. P.,** *Trans. Met. Soc. AIME,* 230, 1035, 1964.
25. **Heydemann, P. L. M.,** *J. Appl. Phys.,* 38, 2640, 1967; *J. Basic Eng.,* 89, 551, 1967.
26. **Zeto, R. J., Vanfleet, H. B., Hryckowian, E., and Bosco, C. D.,** *Accurate Characterization of the High-pressure Environment,* National Bureau of Standards Special Publication 326, U.S. Government Printing Office, Washington, D.C., 1971.
27. **Gschneider, K. A., Elliot, R. O., and McDonald, R. A.,** *J. Phys. Chem. Solids,* 23, 1201, 1962.
28. **Brandt, N. B. and Ginzburg, N. I.,** *Sov. Phys. JETP,* 17, 576, 1963.
29. **Bundy, F. P.,** *J. Appl. Phys.,* 36, 616, 1965.
30. **Darnell, A. J.,** *Bull. Am. Ceram. Soc.,* 44, 634, 1965.
31. **Bridgman, P. W.,** *Proc. Am. Acad. Arts Sci.,* 74, 1, 1940.
32. **Dadson, R. S. and Greig, R. G. P.,** *Br. J. Appl. Phys.,* 16, 1711, 1965.
33. **Bridgman, P. W.,** *Proc. Am. Acad. Arts Sci.,* 81, 167, 1952.
34. **Balchan, A. S. and Drickamer, H. G.,** *Rev. Sci. Instrum.,* 32, 308, 1961.
35. **Bridgman, P. W.,** *Proc. Am. Acad. Arts Sci.,* 74, 425, 1942.
36. **Balchan, A. S.,** Master's thesis, University of Illinois, Urbana, 1959.
37. **Weir, C. E., Van Valkenburg, A., and Lippincott, E.,** in *Modern Very High Pressure Techniques,* Wentorf, R. H., Ed., Butterworths, London, 1962.
38. **Klyuev, Y. A.,** *Sov. Phys. Dokl.,* 7, 422, 1962.
39. **Bassett, W. A., Takahashi, T., and Stook, P. W.,** *Rev. Sci. Instrum.,* 38, 37, 1967.
40. **Jeffery, R. N., Barnett, J. D., Vanfleet, H. B., and Hall, H. T.,** *J. Appl. Phys.,* 37, 3172, 1966.
41. **Jamieson, J. C.,** *Science,* 139, 845, 1963.
42. **Brugger, R. M., Bennion, R. B., and Worlton, T. G.,** *Phys. Lett. A.,* 24, 714, 1967.
43. **Bennion, R. B., Worlton, T. G., Peterson, E. R., and Brugger, R. M.,** unpublished, 1966.
44. **Kennedy, G. C. and La Mori, P. N.,** *J. Geophys. Res.,* 67, 851, 1962.
45. **Cleron, V., Coston, C. J., and Drickamer, H. G.,** *Rev. Sci. Instrum.,* 37, 68, 1966.
46. **Hagelberg, M. P., Holton, G., and Kao, S.,** *J. Acoust. Soc. Am.,* 41, 564, 1967.
47. **Bridgman, P. W.,** *Proc. Am. Acad. Arts Sci.,* 47, 347, 1911.
48. **Zhokhovskii, M. K.,** *Izmer. Tekh.,* 1955, 3, 1955.
49. **Zhokhovskii, M. K., Razuminkhin, V. N., Zolotykh, E. V., and Burova, L. L.,** *Izmer. Tekh.,* 1959, 26, 1959.
50. **Newhall, D. H., Abbot, L. H., and Dunn, R. A.,** *High-pressure Measurement,* Gardini, A. A. and Lloyd, E. C., Eds., Butterworths, London, 1963, 339.
51. **Yasunami, K.,** *Proc. Jap. Acad.,* 43, 310, 1967; *Rev. Phys. Chem. Jap.,* 37, 1, 1967.
52. **Cross, J. L.,** National Bureau of Standards, Washington, D.C., unpublished.
53. **Rossini, F. D.,** *Chemical Thermodynamics,* Wiley and Sons, New York, 1950, chap. 29.

Chapter 7

THE SCALE OF ATOMIC WEIGHTS

A. INTRODUCTORY COMMENTS

In this chapter we summarize the terminology relating to atomic weights, give a historical summary to 1920, discuss the determination of atomic weights by chemical means with analytical experiments and by physical means with the mass spectrometer and nuclear reactions, describe the discovery of the isotopes of oxygen and the resulting two scales of atomic weights, one chemical and one physical, and the eventual unification of these two scales into the Carbon-12 scale of atomic weights.

Most scientists at one time or another deal with molal quantities, either of the actual substances or by way of calculations. In order to deal intelligently and reliably with molal quantities, we need to have knowledge of the scale of atomic weights, its origin and current basis, and the reliability of the numbers presented in the scale. For those scientists and engineers engaged in producing numerical data for science and technology on a molal basis, full knowledge of the scale of atomic weights is imperative.

Following are several simple examples.

To calculate the molal volume of a given substance, under given conditions, one needs to measure the density, d, of the substance under those given conditions, as in g cm^{-3}. The molal volume, V, is obtained from the relation,

$$V = M/d, \tag{1}$$

where M is the molecular weight of the given substance, calculated from the atomic weights of its elements.

In determining the heat (enthalpy) of combustion of methane, one measures the heat evolved, in a suitable reaction vessel in a calorimeter, in the reaction,

$$CH_4\,(g) + 2O_2\,(g) = CO_2\,(g) + 2H_2O(liq), \tag{2}$$

Here the problem is to measure, basically, two quantities: (a) the amount of heat evolved, and (b) the amount of reaction taking place. The amount of reaction is determined from the mass of one of the reactants or one of the products. In this particular example, it happens that the less difficult and more accurate and precise procedure is to determine the mass of carbon dioxide or of water. Each mole of methane produces 1 mol of CO_2 and 2 mol of H_2O. If one has measured the mass of carbon dioxide and divides that mass by the molecular weight of CO_2, one has the number of moles of CH_4. The quantity of energy measured divided by this number of moles gives the heat (enthalpy) of combustion per mole of mathane. It is possible to measure both the mass of carbon dioxide and the mass of water formed in the reaction, with each mole of CH_4 forming 2 mol of H_2O and 1 mol of CO_2. In such a case, if the methane and the oxygen are pure, and if the reaction is "pure," the following stoichiometric relation holds:

$$\frac{\text{mass}\,(CO_2)}{M\,(CO_2)} \bigg/ \frac{\text{mass}\,(H_2O)}{2M\,(H_2O)} = 1 \text{ (exactly)}, \tag{3}$$

Divergence of this number from unity would indicate an error in the assigned values of the atomic weights of H, C, or O. With O = 16, as the base of the former scale of atomic weights, and with oxygen constituting 89% of the molecular weight of H_2O and 73% of the molecular weight of CO_2, it would appear that any real divergence of the ratio from unity would reflect a need for an alteration in the assigned value for the atomic weight of carbon, then based on the oxygen scale.

Other similar examples can be constructed to illustrate the importance of knowing about atomic weights, their reliability, etc.

B. TERMINOLOGY RELATING TO ATOMIC WEIGHTS

Following are some terms and definitions involved in a proper understanding of atomic weights and atomic masses.

The atomic number, Z, of a given atom is the integer giving the number of protons in the nucleus of the atom, which corresponds to the electronic charge on the nucleus and also to the number of electrons outside the nucleus in the neutral atom.

The number of neutrons in the nucleus of a given atom is denoted by N.

The mass number, A, of a given atom is the integer giving the sum of the number of protons and neutrons in the nucleus, or, usually, the total number of nucleons in the nucleus:

$$A = Z + N. \tag{4}$$

For values of Z less than about 20, the values of A are around $2Z$ for the stable nuclei. As Z increases in value, A becomes increasingly greater than $2Z$. For uranium, $Z = 92$, A has values around 235.

Nuclide refers to any given species of atoms, identified by the constitution of its nucleus, with each atom having identical atomic number and identical mass number.

A mononuclidic element is one that consists of only one nuclide.

Two or more nuclides that have the same atomic number, Z, but different mass numbers, A, are called isotopes of the given element.

Normally, for an element consisting of two or more isotopes, when the chemical symbol for that element is written without any designation of mass number, the symbol represents the naturally occurring mixture of isotopes of that element.

The convention for designating the atomic number, Z, the mass number, A, the number and kind of charges on the nuclide, and the number of atoms of the nuclide in the molecule is illustrated for oxygen as follows:

$${}^{16}_{8}O_2^{2+}. \tag{5}$$

Here we have $Z = 8$, $A = 16$, two positive charges on the nuclide, and two atoms constituting the molecule ion.

The atomic weight, or relative atomic mass, of an element is the ratio of the average mass per atom of the natural nuclidic composition of an element to 1/12 of the mass of an atom of the nuclide Carbon-12, ^{12}C.

Strictly, the term "atomic weight," meaning relative atomic mass of the naturally occurring mixture of isotopes of a given element referred to the reference base, should be abandoned, but the International Union of Chemistry has continued its use in order to reserve the designation "atomic mass" for nuclides as distinguished from elements. It is in this arbitrary but useful sense that the term atomic weight is used as a dimensionless number by IUPAC and in this book.

The mole is the amount of a substance, of specified chemical formula, containing the same number of formula units (molecules, atoms, ions, electrons, or other entities) as there are atoms in 12 g (exactly) of the pure nuclide ^{12}C, Carbon-12.

For further details on terminology, the reader is referred to Condon and Odishaw,[1] McGlashan,[2] and Greenwood.[11]

C. HISTORICAL SUMMARY, TO 1920, OF SCALES OF ATOMIC WEIGHTS

Without too much detail, the following constitutes a summary of the history of scales of atomic weights up to the year 1920.[3,4]

In 1803, Dalton (United Kingdom) published a scale of atomic weights based on H = 1. He selected hydrogen as the basis because it has the smallest atomic weight. Avogadro and Cannizaro (both of Italy) accepted Dalton's scale.

In 1818, Berzelius (Sweden) chose the arbitrary base O = 100 for his scale of atomic weights. He reasoned that oxygen was a good base because it combined readily with many more elements than did hydrogen and hence could serve better as a reference standard.

Later, Dalton's scale appeared to be the one most widely used, and it continued to be the preferred scale through the middle of the 19th Century.

In 1883, Meyer and Seubert, using a new method, determined the ratio of the mass of H to be 15.96, and later corrected this value to 15.879. (The present "best" value for this ratio is 15.874.)

In 1900, Clarke (United States) published, as the seventh annual report of the United States Committee on Atomic Weights,[5] two tables of atomic weights, one with H = 1 and one with O = 16.

Meanwhile, in Germany, the German Chemical Society had taken the initiative in dealing with atomic weights on an international scale.[6] Fischer promoted the establishment of a working commission (Landolt, Chairman; Ostwald and Seubert) to contact chemical societies in other countries with the view to setting up an international committee to resolve the confusion surrounding values of atomic weights. In its initial survey, the German Commission received 49 replies from organizations and some individuals regarding their preference as to the hydrogen or oxygen scales of atomic weights. The results were as follows:[7] 40 in favor of O = 16; 7 in favor of H = 1; 2 undecided. About a year later, the German Commission published the results of a second and larger survey, with the following results:[8] 106 in favor of H = 1; 78 in favor of O = 16. However, several eminent German chemists expressed the opinion that it was doubtful that such matters could be decided by a vote of this kind.

The inconclusive results of the foregoing "international" surveys led to the establishment in 1902–1903 of a small, standing International Committee on Atomic Weights, with Clarke (U.S.) as chairman and Thorpe (U.K.), Seubert (Germany), and Moissan (France) as members.

In 1904 this International Committee on Atomic Weights published its first report[9] under its chairman, Clarke, putting forth a compromise in the form of two tables of atomic weights, one based on H = 1 and the other on O = 16. Apparently, this compromise resulted from Clarke's strong preference for the hydrogen scale.

In 1905 this Committee published its second report,[10] again with two tables of atomic weights, based on the hydrogen and oxygen scales. In 1908–1909, it finally agreed on one scale of atomic weights, and published a scale based on O = 16.

In 1911 the International Committee on Atomic Weights became associated with the International Association of Chemical Societies, but, during World War I, this body became inactive.

In 1918 the International Union of Pure and Applied Chemistry (IUPAC) was formed in London. In 1920 IUPAC created a new Commission on Atomic Weights, charged with producing, at appropriate intervals, an internationally acceptable table of atomic weights based on O = 16.

D. DETERMINATION OF ATOMIC WEIGHTS BY CHEMICAL MEANS

Up to about 1920, the determination of the atomic weights of the elements, other than the inert gases, was limited to measurements involving quantitative analytical chemistry. Such measurements, to be useful, were required to be precise and accurate and accomplished with some degree of sophistication. Prominent among the many investigators who devoted some or all of their efforts to this work were the following: J. Berzelius, J. B. A. Dumas, C. de Marignac, J. S. Stas, T. W. Richards, G. B. Baxter, O. Honigschmid, and B. Brauner.

The basic method in such chemical investigations is to determine the precise amount of one element that combines exactly with a given amount of another element of known atomic weight. Following are several examples, based on the former reference base, O = 16.

To determine the atomic weight of hydrogen, referred to 0 = 16, one takes a known mass of hydrogen, and combines it in a "pure" reaction with oxygen to form a measured amount of pure water, according to the reaction:

$$H_2 + \tfrac{1}{2} O_2 = H_2O. \tag{6}$$

With the masses of H_2 and H_2O determined, and the atomic weight of O taken as 16, the atomic weight of H is determined.

To determine the atomic weight of carbon, referred to the former base O = 16, one takes a known mass of pure carbon, and combines it with oxygen to form a measured amount of pure carbon dioxide, according to the reaction:

$$C + O_2 = CO_2. \tag{7}$$

The atomic weights of metals that form known pure oxides are determined by oxidizing a known mass of pure metal, M, to form a measured mass of pure oxide, according to the generalized reaction

$$aM + bO_2 = M_aO_{2b} \tag{8}$$

where a and b are small integers.

In general, having determined X with reference to O, one can proceed in a chain to use reactions not involving the reference base directly, as:

$$\text{OX; XY; YZ; etc.} \tag{9}$$

Historically, one of the most common ratios used was the measurement of the ratio of Cl to AgCl, with Ag having been previously determined by measurement of the ratio of Ag to $AgNO_3$, with N and O known.

Chemical determinations of atomic weights require exacting attention to detail, the preparation of the reacting substances in a degree of highest purity, the careful examination of the chosen reaction to establish its "purity" with respect to absence of side reactions and the purity of the product or products, and the precise determination of the masses of the substances, reactants and products, chosen to determine the atomic weight of the "unknown" element.

Further details on the history of the chemical determination of atomic weights are given by Wichers,[4] Greenwood,[11] and Cameron.[12]

E. DETERMINATION OF ATOMIC WEIGHTS BY THE GAS DENSITY METHOD

For the inert gases, and a few other substances, the atomic weights have been determined by measurements of the density of the selected gases at a series of pressures reduced from atmospheric down to low pressures, limited by the precision of measurement.

The basic principle of such measurements is as follows: At a given constant temperature, maintained precisely, measurements are made, on a given mass of the selected gaseous compound, of the volume of the container and the pressure, at a series of reduced pressures. A plot of the product of pressure x volume against the pressure permits easy extrapolation of the value of the pressure x volume product to zero pressure, as shown in Figure 4.1 of Chapter 4. At zero pressure, the following relation holds,

$$PV = nRT = \frac{m}{M} RT \qquad (10)$$

where n is the number of moles and m is the mass of the sample, M is the molecular weight to be determined, R is the gas constant per mole, T is the absolute temperature, and P and V are the measured pressure and volume, respectively.

By making such measurements in a "substitution" experiment with pure oxygen, it is possible to eliminate a number of constant errors by ending up with the ratio of

$$m/M = m'/M' \qquad (11)$$

where m and m' are measured and M' is the known molecular weight of the pure oxygen. The substitution method eliminates absolute errors in the values of the gas constant, R, the absolute temperature, T, and in the measurement of P and V.

Investigators who did much work in this field have included[3] Lord Rayleigh, E. W. Morley, P. A. Guye, A. Leduc, E. Moles, and G. P. Baxter. It should be emphasized that such measurements are very difficult to make with high precision and accuracy.

F. DETERMINATION OF ATOMIC MASSES AND WEIGHTS WITH THE MASS SPECTROMETER

The mass spectrometer serves to determine with high precision the relative masses of two nuclides of not greatly different masses. It was about 1920 that measurements of atomic weights with the mass spectrometer began. In the period 1920 to 1940, such measurements were about competitive in accuracy with atomic weights determined chemically. But, beginning about 1940, the precision and accuracy of measurements with the mass spectrometer increased far beyond the capabilities of the best quantitative chemical analyses, except in a few cases, and, in recent years, essentially all improved values of atomic weights have been derived from mass spectrometer measurements, with some help from measurements of nuclear reactions.

For an element consisting of only one nuclide, the mass spectrometer provides directly the atomic weight with very high precision.

For an element consisting of two or more isotopes, one needs the following data to calculate the atomic weight: (a) the atomic mass of each of the isotopes comprising the element in its naturally occurring state, and (b) the relative amounts of the isotopes comprising the element. As far as the nuclidic masses are concerned, the precision and accuracy of their determination with the mass spectrometer is far beyond the requirements for calculating atomic weights. But the determination of the relative amounts of the isotopic nuclides comprising a given element is much less precise and serves to limit the value of the atomic weight of that element. The most accurate and precise measurements of the relative abundance of the isotopes of a given element are those made by reference to a known synthetic mixture of the isotopes.

Following is a simple example of the calculation of the atomic weight of an element consisting of two isotopes, using some older data on rubidium as reported by Cameron:[12]

Nuclide	Abundance	Nuclidic mass
^{85}Rb	$(1 - X) = 0.7215 \pm 0.0001$	$A85 = 84.91171 \pm 0.00006$
^{87}Rb	$X = 0.2785 \pm 0.0001$	$A87 = 86.90918 \pm 0.00008$

The atomic weight of rubidium derived from these data is

$$\text{At. wt.} = X(A87) + (1-X)(A85) = (A85) + X[(A87) - (A85)] = 85.4680 \quad (12)$$

The uncertainty in this value can easily be obtained by propagating the errors in Equation 12:

$$\Delta(\text{At. Wt.}) = (\pm 0.00006) + (\pm 0.0001)(2.0) = \pm 0.0002. \quad (13)$$

One sees that the resulting uncertainty is essentially simply the difference in the masses of the two isotopes times the uncertainty in the relative abundance of one of the isotopes. As may be seen, the uncertainties in the nuclidic masses are usually of a lower order of magnitude.

Additional detailed information on the determination of atomic weights and nuclidic masses with the mass spectrometer may be obtained from reports by Cameron[12] and Williams and Duckworth.[13]

G. DETERMINATION OF NUCLIDIC MASSES FROM THE ENERGIES OF NUCLEAR REACTIONS

Since it has become possible to measure quite reliably the energies associated with given nuclear reactions, the difference in the atomic masses of the products of such nuclear reactions over the atomic masses of the reactants can be evaluated by means of the Einstein relation:

$$\Delta E = (\Delta m_0)c^2. \quad (14)$$

Here, ΔE is the increase in energy of the products over the reactions, Δm_0 is the increase in the rest mass of the products over the reactants, and c is the velocity of light.

With ΔE expressed in cal mol^{-1} and Δm_0 in g mol^{-1},

$$\Delta m_0 = 4.6553 \times 10^{-14}\ \Delta E. \quad (15)$$

With ΔE in million electronvolts, meV, and Δm_0 in g mol^{-1},

$$\Delta m_0 = 0.0010735\ \Delta E. \quad (16)$$

It is seen that measurements of ΔE reliable to 1% will provide differences in mass good to 10^{-5} g mol^{-1}.

For example, for the nuclear reaction,

$$^{12}C + {}^{4}He = {}^{2}H + {}^{15}N \quad (17)$$

ΔE has been measured to be 4.97 meV. This means that

$$m(^{2}H) + m(^{15}N) - m(^{12}C) - m(^{4}He) = 0.00534 \text{ g mol}^{-1} \quad (18)$$

and permits the precise evaluation of one of the masses when the other three are known.

The foregoing method, which is most useful in connection with the lighter elements, is described in detail by Williams and Duckworth.[13]

In connection with measurements with the mass spectrometer, one should note that correction for the loss of one or more electrons by the parent atom or molecule to form the atomic ion or molecular ion must be made in calculating the masses appropriately. For example, consider the process:

$$^{12}C = {}^{12}C^{+} + e^{-} \quad (19)$$

12 (exactly) 11.99945 0.00055 g mol^{-1}

Since the electron has a rest mass of 0.00055 g mol^{-1}, the appropriate value of the mass of the carbon ion must be used.

According to the Einstein relation, Equation 14, the value of Δm_0 for the ionization of atomic carbon is, for ΔE in cal mol^{-1},

$$\Delta m_0 = 4.6553 \times 10^{-14}\ \Delta E. \quad (20)$$

For this reaction,

$$\Delta E = 260 \text{ kcal mol}^{-1} \quad (21)$$

so that

$$\Delta m_0 = 1.2 \times 10^{-8} \text{ g mol}^{-1}. \quad (22)$$

This means that the sum of the masses of $^{12}C^{+}$ and e^{-} is greater than that of the neutral ^{12}C atom by 1.2×10^{-8} g mol^{-1}. It follows that the change in mass for the ionization of carbon becomes significant only in the eighth decimal place, or 0.01 ppm, in the atomic mass of Carbon-12, or of its ion. Clearly, this difference is quite negligible for the present purposes.

H. DISCOVERY OF THE ISOTOPES OF OXYGEN

Up to 1929, mass spectrometric observations on oxygen had not been precise enough to detect any isotopes in the then-used reference material for the international scale of atomic weights, oxygen, the reference mass of which was taken as 16.

Early in 1929, Giauque and Johnston published the first account of their discovery of the existence of Oxygen-18, followed shortly by their report of the discovery of the existence of Oxygen-17.[15-17] Their work is a very good example of being most observant of all details in a given investigation and not giving up until all pieces fall properly into place. In a private communication,[18] Giauque has written to the author as follows:

> As you know, I started a program in the 1920's to investigate the relationship between the entropies of gases as measured by low temperature calorimetry and the third law of thermodynamics and as calculated from detailed band spectra and quantum statistics... I started H. L. Johnston on a Ph.D. thesis... on O_2 and N_2. The oxygen work was completed before the summer of 1928 and, using only the strong lines in its band spectrum, as published by Dieke and Babcock, the check was excellent. However, I would not release it because the spectrum was not completely understood... I paced the floor during the fall semester of 1928 until November, when I decided that the weak lines had to be due to an isotope.

Giauque went on to say that there were two major difficulties: one was reconciling all of the band spectra data on oxygen, and the other was the fact that Aston, then the world's leading expert in mass spectrometry, had not yet been able to find any isotopes of oxygen, and, in fact, then considered oxygen to be an ideal reference substance for international atomic weights.

This discovery by Giauque and Johnston had a profound effect on the scale of atomic weights. The precision and accuracy of measurements of relative atomic masses with the mass spectrometer increased to the point where the isotopes of oxygen, ^{16}O, ^{17}O, and ^{18}O, were not only readily identified but also characterized as to relative masses and as to their relative abundance in naturally occurring oxygen.

I. TWO SCALES OF ATOMIC WEIGHTS: CHEMICAL AND PHYSICAL

With increasing precision and accuracy in their measurements, both with mass spectrometers and with nuclear reactions, physicists soon came to use Oxygen-16 as the reference for their scale of atomic masses and the resulting atomic weights. This meant that the chemists were using O = 16 (exactly), while the physicists were using $^{16}O = 16$ (exactly), as their respective references. (Note that O means the naturally occurring mixtures of ^{16}O, ^{17}O, and ^{18}O.)

At first, the existence of the two scales caused little difficulty because the differences appeared to be not too significant for much scientific work. However, it should be noted that a corollary of the two scales of atomic weights is that all constants expressed per mole would have different values on the two scales. Such constants include the Avogadro Constant, N, the Gas Constant, R, and the Faraday Constant, $\mathcal{F}$.

By about 1940, the proportions of the three isotopes constituting naturally occurring oxygen were well established with good accuracy, as follows:

Nuclide	Abundance
^{16}O	0.99759
^{17}O	0.00037
^{18}O	0.00204
Total	1.00000

On the basis of present-day nuclidic masses for ^{16}O, ^{17}O, and ^{18}O, the foregoing would give, with ^{16}O = 16 (exactly), O = 16.004462. Thus, the ratio of the two scales would be 16.004462/16 (exactly) or 1.000279. Because of observed differences in the relative abundance of the different isotopes in naturally occurring oxygen from different sources in the world, the ratio of the two scales was, at that time, set at 1.000275 ± 0.000010.

The values of the Faraday Constant, the Gas Constant, and the Avogadro Constant used by chemists and physicists thus differed by 275 ppm.

For a great deal of scientific work, and for most technological work, this difference was tolerable. But there were being generated a great deal of new data on the physical and thermodynamic properties of pure substances, particularly hydrocarbons, which could not tolerate such a difference, the data being considered reliable to near 100 ppm, compared with the indicated difference of 275 ppm.

J. UNIFICATION OF THE CHEMICAL AND PHYSICAL SCALES OF ATOMIC WEIGHTS

On the chemists' scale, 1 mol, N, atoms of the naturally occurring mixture of ^{16}O, ^{17}O, and ^{18}O, had a mass equal to 16 g (exactly).

It followed that the number of molecules in 1 mol, N, on the physicists' scale would be greater than the number of molecules in 1 mol, *N*, on the chemists' scale by 275 ± 10 ppm.

By about 1950, when there was increasing need for chemists and physicists to communicate with and understand each other, the situation had become so confusing that chemists and physicists alike began to cast about for some way of achieving a unification of the scales of atomic weights used by these two disciplines of science. It was noted that most of the precise numerical data of science existing in the literature, dealing with the properties of pure substances, had been generated by chemists and were expressed in terms of the chemists' mole. One way of describing the situation is to say that the physicists had all the logic on their scale in having a nonvariable reference, ^{16}O =16, whereas the chemists had most of the numerical data in the literature. Clearly, any new side would have to be derived such as not to require revision of the extensive numerical data in the literature that had been generated by chemists. Experts in the latter area agreed that any new scale that differed from the existing chemists' scale by not more than ± 50 ppm would obviate the need for revising extensive data in the literature.

As of about 1950, then, the situation could be summarized as follows, with four possible courses of action:

1. Do nothing, and continue with the two scales of atomic weights differing by 275 ± 10 ppm.
2. Adopt ^{16}O = 16 (exactly), the physicists' scale, and have all the precise numerical molal data of chemistry in the literature revised.
3. Adopt ^{16}O = 15.9956*, so that all the precise numerical molal data of chemistry would remain unchanged, but the precise numerical molal data of physics would be changed.
4. Adopt a new reference involving a change from the chemists' scale of not more than 50 ppm.

*The corresponding value today would be 15.994915.

Many different reference nuclides were considered by many people in searching for a solution to the problem of unifying the two scales of atomic weights. Limiting consideration to nuclides that would require a change from the chemists' scale of not more than ± 50 ppm, the following nuclides were considered, in relation to the existing scales, O = 16 and ^{16}O = 16:

^{17}O, ^{18}O, ^{19}F, ^{15}N, ^{12}C

Table 7.1 shows how these were related to the chemists' and physicists' scales.

Considering the need for having as a reference nuclide one which would permit a maximum of intercomparisons with the mass spectrometer, it soon became evident that Carbon-12 was superior to all others. Carbon-12 was originally suggested independently by A. Ölander and A. O. Nier in 1957. Reports on the advantages and desirability of the Carbon-12 Scale, in comparison with other scales, were issued by Mattauch,[19,20] Kohman, Mattauch, and Wapstra,[21] Kieffer,[22] and Labbauf.[23]

Table 7.1

Relation of Several Possible Reference Nuclides to the Former Chemists' and Physicists' Scales

Reference nuclide	Difference between the given scale and the Chemists' Scale (ppm)
O = 16 (exactly)	0
^{16}O = 16 (exactly)	+275
^{17}O = 17 (exactly)	+8
^{18}O = 18 (exactly)	–4
^{19}F = 19 (exactly)	+42
^{15}N = 15 (exactly)	–50
^{12}C = 12 (exactly)	–43

K. THE CARBON-12 SCALE OF ATOMIC WEIGHTS

With the concurrence of the International Union of Pure and Applied Physics, through its Commission on Nuclidic Masses and Related Constants, the International Union of Pure and Applied Chemistry, at its Conference in 1961, on recommendation of its Commission on Atomic Weights, Edward Wichers, Chairman, approved the Carbon-12 Scale to become effective January 1, 1962.

Among the advantages cited for the Carbon-12 scale for intercomparison of atomic masses with the mass spectrometer are the following:

1. No other element forms molecular ions containing as many atoms of one kind (up to ten or more) as does carbon. Example: C_5^+.
2. Doubly, triply, and quadruply charged atoms of ^{12}C can be paired in doublets with nuclides having masses numbers of 6, 4, and 3.
3. Carbon forms many hydride ions, so that many intermediate masses numbers can be compared directly in the range up to 120.
4. For mass numbers in the range 120 to 240, nuclides can be measured by pairing in doublet their doubly charged ions, as X^{++}, with singly charged ions of $^{12}C_n$ or $^{12}C_nH_m$ fragments.
5. Determination of atomic masses of individual nuclides can be made with the mass spectrometer to six, seven, or even eight significant figures.
6. Improvements in the apparatus, technique, and procedures are leading to greater precision and accuracy in determining relative abundances of isotopes, a necessary component in calculating atomic weights for those elements having two or more isotopes.

Table 7.2 illustrates the pattern of intercomparisons with the Carbon-12 Scale.

Adoption of the unified scale, based on ^{12}C = 12 (exactly), resulted in a lowering by 43 ppm of all the values of atomic weights in the former chemists' scale.

Table 7.2

Scheme of Intercomparison of Ions with the Mass Spectrometer on the Carbon-12 Scale

Ratio of charge/mass	Carbon ions	Other ions
3	$^{12}C^{++++}$	$^{3}M^{+}$
4	$^{12}C^{+++}$	$^{4}M^{+}$
6	$^{12}C^{++}$	$^{6}M^{+}$
12	$^{12}C^{+}$	$^{12}M^{+}$
24	$^{12}C_2^{+}$	$^{24}M^{+}$ or $^{48}M^{++}$
36	$^{12}C_3^{+}$	$^{36}M^{+}$ or $^{72}M^{++}$
48	$^{12}C_4^{+}$	$^{48}M^{+}$ or $^{96}M^{++}$
60	$^{12}C_5^{+}$	$^{60}M^{+}$ or $^{120}M^{++}$
72	$^{12}C_6^{+}$	$^{72}M^{+}$ or $^{144}M^{++}$
84	$^{12}C_7^{+}$	$^{84}M^{+}$ or $^{168}M^{++}$
96	$^{12}C_8^{+}$	$^{96}M^{+}$ or $^{192}M^{++}$
108	$^{12}C_9^{+}$	$^{108}M^{+}$ or $^{216}M^{++}$
120	$^{12}C_{10}^{+}$	$^{120}M^{+}$ or $^{240}M^{++}$

Note: Noting that the carbon ions can have H atoms attached to form molecular C_nH_m ions, the ratio of charge/mass numbers intermediate between the numbers in the first column is possible.

L. DISCUSSION

The development of the scale of atomic weights, beginning in 1803 and continuing to the present time, is a story replete with scientific logic, human preferences, international relations, and the need for compromise in achieving a viable system of scientific communication and understanding.

It appears that the Carbon-12 Scale will be with us for a long time – very likely longer than the 53 years of the international existence of the former chemists' scale, O =16.

For additional details not covered here, the reader is referred to the references. The author is indebted for special information to A. E. Cameron, Oak Ridge National Laboratory, and to William F. Giauque, Professor of Chemistry, University of California – Berkeley.

M. REFERENCES

1. **Condon, E. U. and Odishaw, H.,** *Handbook of Physics,* McGraw-Hill, New York, 1967.
2. *Manual of Symbols and Terminology for Physico-chemical Quantities and Units,* Commission on Symbols, Terminology, and Units, McGlashan, M. L., chairman, International Union of Pure and Applied Chemistry, Butterworths, London, 1970.
3. *Encyclopedia Brittanica,* Vol. 2, 14th ed., Encyclopedia Brittanica, New York, 1932.
4. Report of the Commission on Atomic Weights, Wichers, E., chairman, International Union of Pure and Applied Chemistry, 1961.
5. **Clarke, F. W.,** *J. Am. Chem. Soc.,* 22, 79, 1900.
6. **Fischer, E.,** *Ber. Chem.,* 30, 2955, 1897.
7. **Landolt, H.,** German Commission on Atomic Weights, *Ber. Chem.,* 33, 1847, 1900.
8. **Landolt, H.,** German Commission on Atomic Weights, *Ber. Chem.,* 34, 4353, 1901.
9. **Clarke, F. W.,** International Committee on Atomic Weights, *Ber. Chem.,* 37, 7, 1094.
10. **Clarke, F. W.,** International Committee on Atomic Weights, *Ber. Chem.,* 24R, 143, 1895.
11. International Union of Pure and Applied Chemistry, Commission on Atomic Weights, Greenwood, N. N., chairman, *Pure Appl. Chem.,* 21, 95, 1970.
12. **Cameron, A. E.,** *Anal. Chem.,* 35, 23A, 1963.
13. **Williams, P. and Duckworth, H. E.,** *Sci. Prog.,* 60, 319, 1972.
14. **Rossini, F. D.,** *Chemical Thermodynamics,* John Wiley and Sons, New York, 1950, chap. 14.
15. **Giauque, W. F. and Johnston, H. L.,** *Nature,* 123, 318, 831, 1929.
16. **Giauque, W. F. and Johnston, H. L.,** *J. Am. Chem. Soc.,* 51, 1436, 1929.
17. **Giauque, W. F. and Johnston, H. L.,** *J. Am. Chem. Soc.,* 51, 3528, 1929.
18. **Giauque, W. F.,** private communication, May 1973.
19. **Mattauch, J.,** *Z. Naturforsch.,* 13a, 572, 1958.
20. **Mattauch, J.,** *J. Am. Chem. Soc.,* 80, 4125, 1958.
21. **Kohman, T. P., Mattauch, J., and Wapstra, A. H.,** *Science,* 127, 1431, 1958; *Naturwissenschaften,* 8, 174, 1958.
22. **Kieffer, W. F.,** *J. Chem. Educ.,* 36, 103, 1959.
23. **Labbauf, A.,** *J. Chem. Educ.,* 38, 282, 1962.

Chapter 8

THE INTERNATIONAL ATOMIC WEIGHTS OF 1973

A. INTRODUCTORY COMMENTS

In the preceding chapter it was reported that the International Union of Pure and Applied Chemistry (IUPAC) had established a Commission on Atomic Weights in 1920. This commission has continued to work, albeit interrupted for some years during World War II, and in the past years has issued a report every 2 years at the times of the biennial conferences of IUPAC. After approval by the Council of IUPAC, consisting of delegates from the principal countries, the report is released to the scientific community of the world. During each biennium the members of the Commission on Atomic Weights review and examine all new data, from any source, relating to values of the atomic weights of the elements and the atomic masses of the nuclides comprising them. The Commission is careful to minimize revisions of the values of atomic weights and in each report will recommend changes for only a few elements. No changes are made unless the new data are clearly superior and more reliable than the former data.

In recent years the IUPAC Commission on Atomic Weights has been very careful and helpful in its recommendations, delineating many points that are obvious only to the experts in such matters. The material in this chapter is taken from the reports of the IUPAC Commission on Atomic Weights for 1969, 1971, and 1973, with the actual values of the atomic weights for 1973 taken from their last report.[1-3]

The Titular Members of the IUPAC Commission on Atomic Weights, as recorded in their report for 1973, are as follows: N. N. Greenwood (U. K.), Chairman; H. S. Peiser (U.S.A.), Secretary; P. DeBievre (Belgium); S. Fujiwara (Japan); N. E. Holden (U.S.A.); E. Roth (France). The Associate Members include the following: A. E. Cameron (U.S.A.); G. N. Flerov (U.S.S.R.); R. L. Martin (Australia); N. Saito (Japan); H. G. Thode (Canada); A. H. Wapstra (Netherlands).

B. IMPORTANT GENERAL NOTES CONCERNING VALUES OF ATOMIC WEIGHTS

Several important points should be clearly noted in appraising the reliability of values given for the atomic weights of the chemical elements.

Variations in atomic weights resulting from variations in the isotopic composition of certain elements when taken from different sources in the world include the following, expressed in g mol^{-1}: Boron, ±0.003; Carbon, ±0.00005; Hydrogen, ±0.00001; Silicon, ±0.001; Sulfur, ±0.003; Oxygen, ±0.00016.

Changes in the isotopic composition of certain elements can arise from geological events, from laboratory experimentation, and from industrial processes. For most practical purposes, however, such changes would appear to be not significant, except where fractionation is deliberate.

There are 23 mononuclidic elements, and, consequently, their values are given to ultra high precision. These include the following: Be, F, Na, Al, P, Sc, Mn, Co, As, Y, Nb, Rh, I, Cs, Au, Bi, Pr, Tb, Ho, Tm, Ra, Th, Pa.

Those elements having one predominant isotope (say 99% or greater) will have values of very good precision.

Those elements that have a relatively wide variation in isotopic composition will have values of atomic weights of relatively lesser precision.

A few elements such as Boron, Lithium, and Uranium may have relatively large man-made variations in isotopic abundance.

Certain elements such as Lead, Strontium, and to a lesser extent also Magnesium, Calcium, and Lithium may have anomalous composition in certain geological specimens.

Values of atomic weights for certain elements such as Neptunium, Protactinium, Radium, Technicium, and Thorium are given for the most commonly available long-lived isotope only.

The values of atomic weights given in tables by the IUPAC Commission on Atomic Weights apply to elements as they normally occur in nature. Such elements have a "normal" mixture of isotopes, and the atomic weight is assumed to not differ significantly with samples from different sources because of any of the following reasons: its radiogenic source, its extraterrestrial origin, any artificial fractionation of its isotopes, any artificial nuclear reaction, or a rare geological occurrence.

C. THE 1973 INTERNATIONAL TABLE OF ATOMIC WEIGHTS

The material in this and the following section is derived from the 1973 and 1971 reports of the IUPAC Commission on Atomic Weights.[2,3]

Table 8.1 gives the 1973 International Table of Atomic Weights, as recommended by the IUPAC Commission on Atomic Weights.[3] In the table the elements are arranged in order of atomic number. The last column refers to the footnotes which are applicable to the specified elements.

The values of atomic weights in the fourth column are considered reliable to ±1 unit in the last digit recorded, except for those values having an asterisk, *, which are considered reliable to ±3 units in the last digit recorded.

The values given in parentheses apply to certain radioactive elements whose atomic weights depend significantly on the origin. The value actually given for these elements is the atomic mass number of the isotope of that element having the longest known half-life.

D. NOTES ON THE 1973 INTERNATIONAL TABLE OF ATOMIC WEIGHTS

In addition to the footnotes to Table 8.1, which are applicable to the elements specified, the following notes are helpful in the interpretation of the reliability of the values of atomic weights recommended.[2,3]

The 23 mononuclidic elements are assigned values ranging in reliability from ±0.5 to ±2 ppm.

The ten elements that consist predominantly of one isotope (99% or more) are assigned values ranging in reliability from ±2 to ±100 ppm.

The values of atomic weights given for the natural elements with two or more isotopes apply to these elements as they exist naturally on earth.

The 1973 table includes changes in the values of atomic weights over the values given in the 1971 table for only two elements. These are for Nickel, 58.70 in place of the former value, 58.71*,† and for Rhenium, 186.207 in place of the former value, 186.2. The value for Nickel was

TABLE 8.1

1973 International Atomic Weights; Referred to the Relative Atomic Mass of Carbon-12: ^{12}C = 12 (Exactly)

Atomic number	Name	Symbol	Atomic weight	Footnotes
1	Hydrogen	H	1.0079	b,d
2	Helium	He	4.00260	b,c
3	Lithium	Li	6.941*	c,d,e,g
4	Beryllium	Be	9.01218	a
5	Boron	B	10.81	c,d,e
6	Carbon	C	12.011	b,d
7	Nitrogen	N	14.0067	b,c
8	Oxygen	O	15.9994*	b,c,d
9	Fluorine	F	18.99840	a
10	Neon	Ne	20.179*	c,e
11	Sodium	Na	22.98977	a
12	Magnesium	Mg	24.305	c,g
13	Aluminum	Al	26.98154	a
14	Silicon	Si	28.086*	d
15	Phosphorus	P	30.97376	a
16	Sulfur	S	32.06	d
17	Chlorine	Cl	35.453	c
18	Argon	Ar	39.948*	b,c,d,g
19	Potassium	K	39.098*	
20	Calcium	Ca	40.08	g

† See explanation of asterisk (*) in Section 8.C.

TABLE 8.1 (continued)

1973 International Atomic Weights; Referred to the Relative Atomic Mass of Carbon-12: ^{12}C = 12 (Exactly)

Atomic number	Name	Symbol	Atomic weight	Footnotes
21	Scandium	Sc	44.9559	a
22	Titanium	Ti	47.90*	
23	Vanadium	V	50.9414*	b,c
24	Chromium	Cr	51.996	c
25	Manganese	Mn	54.9380	a
26	Iron	Fe	55.847*	
27	Cobalt	Co	58.9332	a
28	Nickel	Ni	58.70	
29	Copper	Cu	63.546*	c,d
30	Zinc	Zn	65.38	
31	Gallium	Ga	69.72	
32	Germanium	Ge	72.59*	
33	Arsenic	As	74.9216	a
34	Selenium	Se	78.96*	
35	Bromine	Br	79.904	c
36	Krypton	Kr	83.80	e
37	Rubidium	Rb	85.4678*	c
38	Strontium	Sr	87.62	g
39	Yttrium	Y	88.9059	a
40	Zirconium	Zr	91.22	
41	Niobium	Nb	92.9064	a
42	Molybdenum	Mo	95.94*	
43	Technetium	Tc	(97)**	
44	Ruthenium	Ru	101.07*	
45	Rhodium	Rh	102.9055	a
46	Palladium	Pd	106.4	
47	Silver	Ag	107.868	c
48	Cadmium	Cd	112.40	
49	Indium	In	114.82	
50	Tin	Sn	118.69*	
51	Antimony	Sb	121.75*	
52	Tellurium	Te	127.60*	
53	Iodine	I	126.9045	a
54	Xenon	Xe	131.30	e
55	Cesium	Cs	132.9054	a
56	Barium	Ba	137.34*	
57	Lanthanum	La	138.9055*	b
58	Cerium	Ce	140.12	
59	Praseodymium	Pr	140.9077	a
60	Neodymium	Nd	144.24*	
61	Promethium	Pm	(145)**	
62	Samarium	Sm	150.4	
63	Europium	Eu	151.96	
64	Gadolinium	Gd	157.25*	
65	Terbium	Tb	158.9254	a

TABLE 8.1 (continued)

1973 International Atomic Weights; Referred to the Relative Atomic Mass of Carbon-12: ^{12}C = 12 (Exactly)

Atomic number	Name	Symbol	Atomic weight	Footnotes
66	Dysprosium	Dy	162.50*	
67	Holmium	Ho	164.9304	a
68	Erbium	Er	167.26*	
69	Thulium	Tm	168.9342	a
70	Ytterbium	Yb	173.04*	
71	Lutetium	Lu	174.97	
72	Hafnium	Hf	178.49*	
73	Tantalum	Ta	180.9479*	b
74	Wolfram (tungsten)	W	183.85*	
75	Rhenium	Re	186.207	c
76	Osmium	Os	190.2	g
77	Iridium	Ir	192.22*	
78	Platinum	Pt	195.09*	
79	Gold	Au	196.9665	a
80	Mercury	Hg	200.59*	
81	Thallium	Tl	204.37*	
82	Lead	Pb	207.2	d,g
83	Bismuth	Bi	208.9804	a
84	Polonium	Po	(209)**	
85	Astatine	At	(210)**	
86	Radon	Rn	(222)**	
87	Francium	Fr	(223)**	
88	Radium	Ra	226.0254	f,g
89	Actinium	Ac	(227)**	
90	Thorium	Th	232.0381	f,g
91	Protactinium	Pa	231.0359	f
92	Uranium	U	238.029	b,c,e,g
93	Neptunium	Np	237.0482	f
94	Plutonium	Pu	(244)**	
95	Americium	Am	(243)**	
96	Curium	Cm	(247)**	
97	Berkelium	Bk	(247)**	
98	Californium	Cf	(251)**	
99	Einsteinium	Es	(254)**	
100	Fermium	Fm	(257)**	
101	Mendelevium	Md	(258)**	
102	Nobelium	No	(255)**	
103	Lawrencium	Lr	–	

*Considered reliable to ±3 units in the last digit recorded.

**The values given in parentheses apply to certain radioactive elements whose atomic weights depend significantly on the origin. The value actually given for these elements is the atomic mass number of the isotope of that element having the longest known half-life.

[a]Element with only one stable nuclide.

[b]Element with one predominant isotope (about 99 to 100% abundance);

variations in the isotopic composition or errors in its determination have a correspondingly small effect on the value of the atomic weight.
[c]Element for which the value of the atomic weight derives its reliability from calibrated measurements (i.e., from comparisons with synthetic mixtures of known isotopic composition).
[d]Element for which known variations in isotopic composition in terrestrial material prevent a more precise atomic weight being given; the atomic weight values should be applicable to any "normal" material.
[e]Element for which substantial variations in the atomic weight from the value given can occur in commercially available material because of inadvertent or undisclosed change of isotopic composition.
[f]Element for which the value of the atomic weight is that of the most commonly available long-lived isotope.
[g]Element for which geological specimens are known in which the element has an anomalous isotopic composition.

From Commission on Atomic Weights, Greenwood, N. N., chairman, International Union of Pure and Applied Chemistry, *Pure Appl. Chem.*, 37, 4, 1974. With permission.

changed by 1 in 5870, and made more reliable by a factor of three. The value for Rhenium was made more reliable by a factor of 100. The reasons for these changes are given in detail in the report of the IUPAC Commission on Atomic Weights.[3]

The 1973 table reverts to the former practice of the 1955 table in including, for certain radioactive elements, the mass number of the isotope of longest half-life, instead of leaving the space blank. The practice, after the 1955 table and through the 1971 table, of not giving any value for these elements was based on the following reasoning: the use of the mass number is too imprecise for much analytical work; users of these elements generally know enough about the source of their material to deduce the best applicable value; new determinations of the half-life; the isotope of longest life is not always the one most widely available, as for Technicium, Plutonium, and Californium. It was because of the desire of many users that the former practice has been reinstated.

For ease in printing and in computer applications, another change made over the 1971 table is the use in the 1973 table of an asterisk, *, to indicate a reliability of ±3 in the last digit recorded. The 1971 table used a subscript small number for this purpose.

The increased use of the separated stable isotopes of the rare gas elements is leading to isotopically depleted material reaching investigators. This is important in the case of Neon, Krypton, and Xenon, but less so in the case of Argon. The greatly increased use of fission-product materials and separated or enriched isotopic materials has prompted the Commission to include Footnote "e" in its 1973 table for Lithium, Boron, Uranium, Neon, Krypton, and Xenon. Similar precaution may be required in the future on the values for Hydrogen, Helium, Carbon, Nitrogen, and Oxygen.

The Commission calls particular attention to Footnote "g" in its 1973 table relating to certain radioactive or radioactive-related elements such as Osmium, Thorium, Uranium, Lithium, Magnesium, Argon, Calcium, Strontium, Lead, and Radium.

The Commission also calls attention to the problem that modern analytical measurements, physical, chemical, or both, yield values of atomic weights for specified samples of elements (other than the mononuclidic one or those having one predominant isotope) much more precise than the variablility due to differences in isotopic composition arising from man-made processes or from differences in natural material from different sources. For this reason, the values of the atomic weights for the elements mentioned in the preceding paragraph are given to a precision inferior to what is actually easily attainable with modern apparatus and techniques.

E. THE 1973 INTERNATIONAL TABLE OF RELATIVE ATOMIC MASSES OF SELECTED NUCLIDES

The material in this section is derived from the 1973 and 1971 reports of the IUPAC Commission on Atomic Weights.[2,3]

Table 8.2 gives the 1973 International Table of Relative Atomic Masses of Selected Nuclides, as recommended by the IUPAC Commission on

TABLE 8.2

1973 International Relative Atomic Masses of Selected Nuclides

Name	Symbol	Atomic number	Mass number	Relative atomic mass[13]	Half-life[14] *
Hydrogen	H	1	1	1.007825	
(Deuterium)	(D)	1	2	2.014102	
(Tritium)	(T)	1	3	3.016049	12.33 a
Helium	He	2	3	3.016029	
			4	4.00260	
Lithium	Li	3	6	6.01512	
			7	7.01600	
Boron	B	5	10	10.01294	
			11	11.00931	
Carbon	C	6	12	12 exactly	
			13	13.003355	
			14	14.00324	5.73×10^3 a
Nitrogen	N	7	14	14.003074	
			15	15.00011	
Oxygen	O	8	16	15.994915	
			17	16.999133	
			18	17.99916	
Neon	Ne	10	20	19.99244	
			21	20.99385	
			22	21.99138	
Magnesium	Mg	12	24	23.98504	
			25	24.98584	
			26	25.98259	
Silicon	Si	14	28	27.97693	
			29	28.97650	
			30	29.97377	
Sulfur	S	16	32	31.97207	
			33	32.97146	
			34	33.96787	
			36	35.96708	
Argon	Ar	18	36	35.96755	
			38	37.96273	
			40	39.96238	
Calcium	Ca	20	40	39.96259	
			42	41.9586	
			43	42.9588	
			44	43.9555	
			46	45.9537	
			48	47.9525	
Copper	Cu	29	63	62.9296	
			65	64.9278	
Krypton	Kr	36	78	77.9204	
			80	79.9164	
			82	81.9135	
			83	82.9141	
			84	83.9115	
			86	85.9106	
Strontium	Sr	38	84	83.9134	
			86	85.9093	
			87	86.9089	
			88	87.9056	
Technetium	Tc	43	97	96.9064	2.6×10^6 a
			99	98.9062	2.13×10^5 a

TABLE 8.2 (continued)

1973 International Relative Atomic Masses of Selected Nuclides

Name	Symbol	Atomic number	Mass number	Relative atomic mass[13]	Half-life[14]*
Xenon	Xe	54	124	123.9061	
			126	125.9043	
			128	127.9035	
			129	128.9048	
			130	129.9035	
			131	130.9051	
			132	131.9041	
			134	133.9054	
			136	135.9072	
Promethium	Pm	61	145	144.9128	18 a
			147	146.9152	2.6234 a
Osmium	Os	76	184	183.9526	
			186	185.9539	
			187	186.9558	
			188	187.9559	
			188	187.9559	
			189	188.9582	
			190	189.9585	
			192	191.9615	
Lead	Pb	82	204	203.9730	
			206	205.9745	
			207	206.9759	
			208	207.9766	
Polonium	Po	84	209	208.9824	102 a
			210	209.9829	138.38 d
Astatine	At	85	210	209.987	8.1 hr
Radon	Rn	86	222	222.0176	3.824 d
Francium	Fr	87	223	223.0197	22 min
Radium	Ra	88	226	226.0254	1.60×10^{3} a
Actinium	Ac	89	227	227.0278	21.77 a
Thorium	Th	90	230	230.0331	7.7×10^{4} a
			232	232.0381	1.40×10^{10} a
Protactinium	Pa	91	231	231.0359	3.25×10^{4} a
Uranium	U	92	233	233.0397	1.58×10^{5} a
			234	234.0409	2.44×10^{5} a
			235	235.0439	7.04×10^{8} a
			236	236.0456	2.34×10^{7} a
			238	238.0508	4.47×10^{9} a
Neptunium	Np	93	237	237.0482	2.14×10^{6} a
Plutonium	Pu	94	238	238.0496	87.8 a
			239	239.0522	2.439×10^{4} a
			240	240.0538	6.54×10^{3} a
			241	241.0568	15 a
			242	242.0587	3.87×10^{5} a
			244	244.0642	8.3×10^{7} a
Americium	Am	95	241	241.0568	433 a
			243	243.0614	7.37×10^{3} a
Curium	Cm	96	242	242.0588	163 d
			243	243.0614	28 a
			244	244.0627	18.1 a
			245	245.0655	8.5×10^{3} a
			246	246.0672	4.76×10^{3} a
			247	247.0703	1.54×10^{7} a

TABLE 8.2 (continued)

1973 International Relative Atomic Masses of Selected Nuclides

Name	Symbol	Atomic number	Mass number	Relative atomic mass[1 3]	Half-life[1 4] *
Curium (continued)			248	248.0723	3.5×10^5 a
			250	250.0784	1.1×10^3 a
Berkelium	Bk	97	247	247.0703	1.4×10^4 a
			249	249.0750	311 d
Californium	Cf	98	251	251.0796	900 a
			252	252.0816	2.63 a
			254	254.0874	6×10 d
Einsteinium	Es	99	253	253.0848	20.47 d
			254	254.0880	276 d
Fermium	Fm	100	257	257.0951	100.5 d

*a = year; d = day; hr = hour; min = minute.

From Commission on Atomic Weights, Greenwood, N. N., chairman, International Union of Pure and Applied Chemistry, *Pure Appl. Chem.*, 37, 4, 1974. With permission.

Atomic Weights.[3] In this table the elements are arranged in order of atomic number and mass number. The values given are believed reliable to ±1 unit in the last digit recorded.

For the values of the relative atomic masses given in Table 8.2, the Commission refers to Wapstra[4] and Wapstra and Gore[5] and, for the values given for the half-life, to Holden and Walker.[6]

F. DISCUSSION

As may be seen from the 1973 report of the IUPAC Commission on Atomic Weights the advances made in the precision and accuracy of physical and, in some cases, chemical, measurements of atomic weights have far outstripped the constancy of the composition of many chemical elements as they come into the hands of scientific investigators from natural or man-made sources. This new phenomenon produces problems of which investigators must be aware. Appropriate precautions are required to be made.

It is also easy to comment that for 33 of the chemical elements the values of the atomic weights are reliable to ±0.5 ppm for one up to ±100 ppm for the least well-known value in this group. This constitutes a most significant advance in man's ability to measure natural things.

The author is indebted for special information to N. N. Greenwood, Chairman, and H. S. Peiser, Secretary, of the Commission on Atomic Weights, International Union of Pure and Applied Chemistry.

G. REFERENCES

1. Commission on Atomic Weights, Greenwood, N. N., chairman, International Union of Pure and Applied Chemistry, *Pure Appl. Chem.*, 21, 95, 1970.
2. Commission on Atomic Weights, Greenwood, N. N., chairman, International Union of Pure and Applied Chemistry, *Pure Appl. Chem.*, 30, 639, 1972.
3. Commission on Atomic Weights, Greenwood, N. N., chairman, International Union of Pure and Applied Chemistry, *Pure Appl. Chem.*, 37, 4 (1974).
4. **Wapstra, A. H.**, private communication to the IUPAC Commission on Atomic Weights (1973).
5. **Wapstra, A. H. and Gore, N. B.**, Atomic Mass Table, *Nucl. Data Tables*, 9, 265, 1971.
6. **Holden, N. E. and Walker, F. W.**, Chart of Nuclides, Educational Relations, General Electric Company, Schenectady, New York, 1972, 12305.

Chapter 9

THE MECHANICAL EQUIVALENT OF HEAT; "WET" CALORIES

A. INTRODUCTORY COMMENTS

About 200 years ago, little was known about heat, heat capacity, and latent heat. The phenomenon of heat energy was somewhat of a mystery. Two scientists of that time did much to clarify matters, Joseph Black (Scotland) and John Wilcke (Sweden).

Black, who was Professor of Medicine and Lecturer in Chemistry at the University of Glasgow, 1756 to 1766, and Professor of Medicine and Chemistry at the University of Edinburgh, 1766 to 1799, recognized the importance of the thermometer and distinguished between the intensity and quantity of heat. Illustrative of his early work was the following: he found that 1 lb of water at 52°F melted the same quantity of ice at 32°F as did 2 lb of water at 42°F. From a large number of experiments, Black concluded that different substances have different capacities for heat, and measured the heat capacity of many substances. Black observed that if he added heat to ice at 32°F, the temperature of the resulting mixture of ice and water remained constant until all of the ice was melted. He made corresponding observations on adding heat to liquid water at its boiling point. Included among the data obtained by Black were values for the heat of fusion of ice and the heat of vaporization of water, using units related to the heat capacity of liquid water.

Wilcke, who was Professor of Experimental Physics at the Military Academy in Stockholm and also Secretary of the Swedish Academy, carried on his investigations about the same time as, or a little later than, Black. Wilcke also produced data on the heat capacity of many substances, referred to the heat capacity of water.

Near the end of the 18th Century, Count Rumford made his contribution to this science, when, as a result of experimentation, he concluded that the heat produced from the boring of canon came from friction, with the mechanical energy of boring being converted to heat. He reported these observations to the Royal Society (London) in 1798.

By the year 1800 the phenomena of heat capacity and latent heat were reasonably well clarified, as well as the ideas relating to the conversion of mechanical energy into heat energy.

B. EARLY UNIT OF ENERGY

Beginning in the early part of the 19th Century, it became the practice to measure heat energy in terms of the increase in temperature of a given mass of water. For most scientific work the unit generally used was the calorie, taken as the quantity of energy required to raise the temperature of 1 g of water through 1°C. However, in the United Kingdom and the United States, practical engineers tended to use the British thermal unit, defined as the quantity of energy required to raise the temperature of 1 lb of water through 1°F.

The quantity of heat energy to be measured was introduced into a measured mass of water, m, and the resulting rise of temperature, Δt, was measured with a suitable thermometer. The quantity of heat energy, Q, then was given by the relation:

$$Q = (m)\,(\Delta t). \qquad (1)$$

However, it is easy to see that this equation is an approximation because it does not take into account either the heat capacity of the container or that of the thermometer. To improve the accuracy of his work, the investigator then determined the "water equivalent" of the container and the thermometer. Then it became clear that stirring would improve the process of equilibration of the heat energy throughout the system, and the apparatus was modified by including a stirrer. Determination was then made of the "water equivalent" of the stirrer, and, separately, of the energy introduced by the stirring. Finally, it became clear that it was necessary to take cognizance of the heat leak, or thermal flow, between the container and its surroundings. To improve this aspect of the experimentation, a fixed environment was provided with a cover, with holes for the passage of the thermometer and stirrer.

About 1880, Rowland[3] showed that the heat capacity of water varied with the temperature, actually decreasing from 5 to 30°C. We know now that liquid water has a minimum of heat capacity at a temperature near 35°C, with the value at 0°C being higher by 0.95%, and that at 100°C higher by 0.90%. This variation of nearly 1% was recognized as being far beyond the then-current limits of measurement.

C. VARIOUS CALORIES

Following the observation of the variation in the heat capacity of water with temperature, coupled with improvements in the precision of measurement, investigators began to report their results in terms of the heat capacity of water at the mean temperature of the experiments. This gave rise gradually to the appearance in the literature of various calories, including the 4° calorie, the 15° calorie, the 18° calorie, the 20° calorie, and the mean (0 to 100°) calorie.

By about 1900, experimental calorimetry had advanced to a stage where measurements of heat energy, in terms of the heat capacity of water, could be made with a precision of about 1 part in 1,000, sometimes better. In order to compare measurements reported in terms of different calories, it became necessary to investigate the heat capacity of water as a function of temperature.

D. THE MECHANICAL EQUIVALENT OF HEAT

It was early recognized, however, that notwithstanding the relative ease with which measurements of heat energy could be made in terms of the heat capacity of water, it was necessary to ascertain what quantity of energy in absolute units, as ergs, a given calorie was equivalent to.

This gave rise to the series of investigations on the mechanical equivalent of heat begun by Joule about 1840 and continued by him for nearly a half century.[4]

In 1850 Joule reported that 1 Btu-60°F was equivalent to 772 ft-lb. About this same time some electrical equipment for measuring power became available, and with this apparatus Joule obtained, with electrical energy, a value about 1% lower. But it was clear at that time that the accuracy of the measurement of the mechanical equivalent was superior to that of the electrical equivalent.

In 1878 Joule reported a new value for the mechanical equivalent of heat, with 1 Btu-60°F equal to 772.55 ft-lb. Also, about this same time Joule discovered that the electrical resistance previously used in the measurement of electrical power was found to be 1.3% lower in value.

In 1880 Rowland[3] reported the results of a new determination of the mechanical equivalent of heat, made with a very elaborate and improved apparatus. He found that one 20° calorie was equal to 4.179 J.

In 1897 Reynolds and Moorby[5] reported the results of their experiments on the mechanical equivalent of heat. They found that one mean (0 to 100°) calorie was equal to 4.184 J.

Reviews of the then-existing data on the mechanical equivalent of heat were published by Ames[6] in 1900 and by Barnes[7] in 1904.

In 1910 Rispail[8] reported the results of his measurements of the mechanical equivalent of heat.

In 1927 Laby and Hercus[9] published the results of their work on the mechanical equivalent of heat.

The results of the several foregoing investigations on the mechanical equivalent of heat were reviewed and recalculated by Laby and Hercus,[10] who gave the following summary of them in joules equivalent to the 20° calorie: Rowland, 4.182; Reynolds and Moorby, 4.1813*; Rispail, 4.180; Laby and Hercus, 4.1809. (The present best value, based on electrical measurements, is near 4.1818, indicating that these investigators had indeed done very good work for those days.)

E. DISCUSSION

The calories that represent the actual real heat capacity of water at a specified temperature have come to be known as "wet" calories to distinguish them from the artificial, defined, or "dry" calorie discussed in the following chapter.

In this chapter we have depicted the development of the several "wet" calories as units of energy, with some indication of their relation to the fundamental unit of energy. Details not given here will be found in the references. The author gives a summary discussion of the problem.[11]

*Laby and Hercus gave 4.1836 for the mean calorie from Reynolds and Moorby. The author has converted this to the 20° calorie by dividing by 1.00054.

F. REFERENCES

1. Encyclopaedia Britannica, Vol. 4, 11, and 19, 14th ed., Encyclopaedia Britannica, New York, 1932.
2. **McKie, D. and Heathcote, W. H. De V.**, *The Discovery of Specific and Latent Heats,* Arnold and Co., London, 1935.
3. **Rowland, H. A.**, *Proc. Am. Acad. Arts Sci.,* 15, 75, 1880.
4. **Joule, J. P.**, *Phil. Mag.,* 23, 263, 347, 435, 1843.
5. **Reynolds, O. and Moorby, W. H.**, *Trans. R. Soc.* (London), A190, 301, 1897.
6. **Ames, J. S.**, *Rapports Congress International de Physique,* Vol. 1, Gauthier-Villars, 1900, 178, 214.
7. **Barnes, H. T.**, *Transactions of the International Electrical Congress,* Vol. 1, St. Louis, 1904, 53.
8. **Rispail, M. L.**, *Ann. Chim. Phys.,* 20, 417, 1910.
9. **Laby, T. H. and Hercus, E. O.**, *Trans. R. Soc.* (London), A227, 63, 1927.
10. **Laby, T. H. and Hercus, E. O.**, in *International Critical Tables,* Vol. 5, Washburn, E. W., Ed., McGraw-Hill, New York, 1929, 78.
11. **Rossini, F. D.**, *Chemical Thermodynamics,* John Wiley and Sons, New York, 1950, chap. 5.

Chapter 10

THE MODERN UNIT OF ENERGY; "DRY" CALORIES; THE JOULE

A. INTRODUCTORY COMMENTS

As mentioned in the preceding chapter, electrical measurements of energy began to be possible in a practical way before the middle of the 19th Century, but the standards upon which such measurements were based had not then yet been properly and fully developed to provide the accuracy required for basic scientific measurements. Contributions to the work of developing suitable electrical standards were made by a number of individual scientists and by several scientific organizations:[1] Jacobs in 1848; Werner von Siemens in 1860; the British Association for the Advancement of Science (BAAS) in 1861; Gauss; Weber; Clark in 1872; Rowland in 1878. During this time, the Siemens unit of resistance (a column of mercury 1 m in length with a uniform cross section of 1 mm^2, maintained at 0°C) was used in Europe and the BAAS unit (resistance wire, two parts by weight of silver to one part by weight of platinum) was used in the English-speaking countries.

B. DEVELOPMENT OF THE INTERNATIONAL ELECTRICAL UNITS

Reference 1 gives a complete history of the development of the electrical standards, from which the following is extracted.

In 1881, an "International Congress of Electricians" was held in Paris, at which the following resolutions were agreed upon:

1. That the cgs system of electromagnetic units be adopted as the fundamental units;
2. That the practical units, the ohm and the volt, preserve their previous definitions, 10^9 and 10^8 cgs units, respectively;
3. That the unit of resistance, the ohm, be represented by a column of mercury 1 mm^2 in cross section at the temperature of 0°C;
4. That an International Commission be charged with the determination, by new experiments, of the length of the mercury column 1 mm^2 in cross section at a temperature of 0°C;
5. That the current produced by 1 volt in 1 ohm be called an ampere (A);
6. That the quantity of electricity produced by a current of 1 ampere in 1 second be called a coulomb;
7. That the unit of capacity be called a farad which is defined by the condition that 1 coulomb in 1 farad raises the potential 1 volt.

The International Commission referred to in item 4 above had its recommendations approved at an International Conference in 1884, in the form of the following resolutions:

1. The legal ohm is the resistance of a column of mercury 1 mm^2 in cross section and 106 cm in length at the temperature of melting ice;
2. The ampere is equal to one tenth of a cgs unit (of current) of the electromagnetic system;
3. The volt is the electromotive force that will maintain a current of 1 ampere in a conductor of which the resistance is 1 legal ohm.

The value adopted for the length of the mercury column (1 above) was taken as 106 cm, in spite of the fact that the best results were very close to 106.3 cm. Conservatively, the value was rounded to the nearest unit centimeter, even though this might involve an error of 0.3%. On account of this uncertainty, no steps were taken to have this "legal ohm" legalized by the several governments represented.

In 1890 a second "International Congress of Electricians" was held in Paris, which included the approval of the following resolutions, among others:

1. The joule, the practical unit of energy, is equal to 10^7 cgs units.
2. The practical unit of power is the watt, equal to 10^7 cgs units, or 1 joule/sec.

At this stage of development, increased accuracy was obtainable with apparatus of improved construction, and this led to better accord among the various redeterminations of the absolute electrical units in the several countries.

In 1892, at a meeting of the British Association

for the Advancement of Science in Edinburgh, at which representatives of Germany, France, and the United States were present, agreement was reached on new, improved specifications for the ohm, the ampere, and the volt. Their recommendations were put before the International Electrical Congress in Chicago in 1893, which approved them in the following resolutions:[1]

Resolved, That the several governments represented by the delegates of this International Congress of Electricians be, and they are hereby, recommended to formally adopt as legal units of electrical measure the following:

Ohm – As a unit of resistance, the *international ohm,* which is based upon the ohm equal to 10^9 units of resistance of the cgs system of electromagnetic units, and is represented by the resistance offered to an unvarying electric current by a column of mercury at the temperature of melting ice 14.4521 grams in mass, of a constant cross-sectional area and of the length of 106.3 cms.

Ampere – As a unit of current, the *international ampere,* which is one tenth of the unit of current of the cgs system of electromagnetic units, and which is represented sufficiently well for practical use by the unvarying current which, when passed through a solution of nitrate of silver in water, and in accordance with accompanying specifications, deposits silver at the rate of 0.001118 of a gram per second.

Volt – As a unit of electromotive force, the *international volt,* which is the electromotive force that, steadily applied to a conductor whose resistance is 1 international ohm, will produce a current of 1 international ampere, and which is represented sufficiently well for practical use by $\frac{1000}{1434}$ of the electromotive force between the poles or electrodes of the voltaic cell known as Clark's cell, at a temperature of 15°C, and prepared in the manner described in the accompanying specifications.

Coulomb – As a unit of quantity, the *international coulomb,* which is the quantity of electricity transferred by a current of 1 international ampere in one second.

Farad – As a unit of quantity, the *international farad,* which is the capacity of a condenser charged to a potential of 1 international volt by 1 international coulomb of electricity.

Joule – As a unit of work, the *joule,* which is equal to 10^7 units of work in the cgs system, and which is represented sufficiently well for practical use by the energy expended in one second by an international ampere in an international ohm.

Watt – As a unit of power, the *watt,* which is equal to 10^7 units of power in the cgs system, and which is represented sufficiently well for practical use by work done at the rate of 1 joule per second.

Henry – As the unit of induction, the *henry,* which is the induction in a circuit when the electromotive force induced in this circuit is 1 international volt, while the inducing current varies at the rate of 1 ampere per second.*

Although the resolutions of the 1893 Chicago Congress were adopted practically unanimously, only six of the ten governments represented legislated on the subject. These six were the United States, the United Kingdom, Canada, Germany, France, and Austria. Strictly speaking, no two countries defined the electrical units in the same way, undoubtedly because of the incompleteness of the definitions of the 1893 Chicago Congress, as follows:

1. Only two of the three units, ohm, ampere, and volt, should be defined in terms of concrete standards, as the third is derived from the other two;
2. The internationally agreed upon units were not clearly recognized as separate and distinct from the absolute units on which they were based;
3. The specifications for the silver voltameter were inadequate;
4. Redeterminations of the electromotive force of the Clark cell at 15°C indicated a value nearer 1.433 than 1.434 absolute volts.

It should be noted that in some countries the law defined the same unit in three or four different ways.

Then followed international congresses or conferences in Paris in 1900, in St. Louis in 1904, and in Berlin in 1905. At the 1905 Berlin Conference, representatives were present from Austria, Belgium, the U.K., France, Germany, and the U.S., and recommendations were formulated, including such as the following: only two electrical units shall be selected as standards, the third to be derived; the two standards selected are the international ohm and the international ampere; the saturated cadmium (Weston) cell is to replace the Clark cell.

At the International Conference on Electrical Units and Standards held in London in 1908, delegates from 21 countries were present. The following definitions were agreed upon:[1]

The fundamental units are

1. The *ohm,* the unit of electric resistance, which has the value of 1,000,000,000 (10^9) in terms of the centimeter and the second;
2. The *ampere,* the unit of electric current, which

*From *Electrical Units and Standards,* National Bureau of Standards Circular 60, U.S. Government Printing Office, Washington, D.C., 1920.

has the value of one tenth (0.1) in terms of the centimeter, gram, and second;

3. The *volt,* the unit of electromotive force, which has the value of 100,000,000 (10^8) in terms of the centimeter, gram, and second;

4. The *watt,* the unit of power, which has the value of 10,000,000 (10^7) in terms of the centimeter, the gram, and the second.

As a system of units representing the above and sufficiently near for the purposes of electrical measurements, and as a basis for legislation, the conference recommended the adoption of the international ohm, the international ampere, the international volt, and the international watt, defined as follows:

1. The *international ohm* is the resistance offered to an unvarying electric current by a column of mercury at the temperature of melting ice, 14.4521 grams in mass, of a constant cross-sectional area and of a length of 106.300 centimeters.

2. The *international ampere* is the unvarying electric current which, when passed through a solution of nitrate of silver in water, in accordance with the Specification II attached to these Resolutions, deposits silver at the rate of 0.00111800 of a gram per second.

3. The *international volt* is the electrical pressure which, when steadily applied to a conductor the resistance of which is one international ohm, will produce a current of one international ampere.

4. The *international watt* is the energy expended per second by an unvarying electric current of one international ampere under an electric pressure of one international volt.*

It was not strictly logical to specify the precise amount of silver deposited by an ampere because the specifications of the apparatus were not fully adequate. However, subsequent work showed that the number was fortunately chosen and was much nearer the absolute value than the then-current best data indicated.

It then remained to adopt a value for the saturated cadmium (Weston) cell, to supersede the tentative value of 1.084 international volts at 20°C. Because the specifications were not fully adequate, an International Commission was established to determine a "best" value for the Weston cell at 20°C. In 1910 representatives of the National Bureau of Standards (U.S.), the Physikalische Technische Reichsanstalt (Germany), the National Physical Laboratory (U.K.), and the Laboratoire Central d'Electricité (France) gathered at the National Bureau of Standards in Washington and performed extensive comparisons of their respective resistance standards and Weston cells. They found that, in terms of the international ohm and international ampere, as defined at the 1908 Conference, the electromotive force of the saturated cadmium (Weston) cell at 20°C was 1.0183 volts, within ±0.01%. Expressed as international volts, this number was to be taken as exact until redefined. This value for the standard Weston cell became effective January 1, 1911, and was assigned to the mean of the groups of cells that had been brought together for comparison in Washington the preceding year from the U.S., the U.K., France, and Germany.

The international ohm, international ampere, and international volt thus emanating from the 1908 International Conference in London continued to be the international electrical standards in the principal countries until full conversion to the absolute electrical standards came about.

At the time of fixing the international electrical standards of 1908, the selected standards were equal to the absolute measures within the limits of uncertainty of those determinations at that time. However, subsequent work carried on in the several national standardizing laboratories indicated definitely that the international ohm and volt differed from the absolute ohm and volt by quite significant amounts. By about 1920 such comparisons indicated that the international ohm was too large by about 0.05% and the international volt too large by about 0.04%. In terms of electrical energy, these differences would make the international joule greater than the absolute joule by about 0.03%.

In 1930 the National Bureau of Standards reported that 1 international joule was equal to 1.00036 absolute joule. This figure, rounded to the nearest part in 10,000, became 1.0004, as recommended by the National Bureau of Standards.[10]

In 1939 the National Bureau of Standards revised its recommendations significantly, such that 1 international joule was equal to 1.00020 ± 0.00005 absolute joule.[11]

In 1947, following the completion of new and more elaborate measurements comparing the 1908 international electrical standards with the corresponding absolute units, the National Bureau of Standards reported results that made 1 international joule equal to 1.000165 ± 0.00025 absolute joule.[12]

*From *Electrical Units and Standards,* National Bureau of Standards Circular 60, U.S. Government Printing Office, Washington, D.C., 1920.

The foregoing results are summarized chronologically as follows, in terms of the number of absolute joules equal to 1 international J based on the 1908 international electrical standards:

	1 international joule =	
1908	1.0000	absolute joules
1920	1.00034	absolute joules
1930	1.0004	absolute joules
1939	1.00020 ± 0.00005	absolute joules
1947	1.000165 ± 0.000025	absolute joules

The foregoing figures, while applying specifically to the National Bureau of Standards, represent essentially the results obtained at about the same dates in the national standardizing laboratories of other countries.

All measurements of electrical power made from about 1910 to 1948 by means of standard cells and resistances are actually in terms of international watts; hence, the unit of energy was international joules. Beginning January 1, 1948, the National Bureau of Standards, along with the national standardizing laboratories of other countries, has certified standard cells and resistances in absolute volts and ohms, so that the resulting energy is measured in absolute joules.[12]

C. CALORIMETRIC MEASUREMENTS WITH ELECTRICAL ENERGY

Mention has been made in the previous chapter of the fact that Joule did a few measurements with electrical energy in 1850 and in 1878, when the electrical standards were not yet adequate for accurate measurements. He found that, at those times, the mechanical method was more reliable than the electrical method. However, important improvements continued to be made in the development and maintenance of electrical standards, as reported in the preceding section of this chapter.

After Joule, measurements of the "mechanical equivalent of heat," or more properly the heat capacity of water, with electrical energy were made by a number of investigators up to 1915, as follows: Griffiths[2] in 1894; Schuster and Gannon[3] in 1895; Callendar and Barnes[4] in 1909; Bousfield and Bousfield[5] in 1911; and Jaeger and Steinwehr[6] in 1915. The results of these investigations were recalculated by Laby and Hercus[7] in terms of the international electrical units of 1908 and the 20° calorie with the following results, in international joules deg^{-1} g^{-1}: Griffiths, 4.1904; Schuster and Gannon, 4.1898; Callendar and Barnes, 4.1795; Bousfield and Bousfield, 4.1767; Jaeger and Steinwehr, 4.1821.

In 1939 definitive measurements of the heat capacity of water were made at the National Bureau of Standards by Osborne, Stimson, and Ginnings,[8] superseding the earlier measurements in the same laboratory by Osborne, Stimson, and Fiock.[9] These measurements were made with electrical energy in terms of international joules based on the 1908 international electrical standards. The results of this investigation have been recalculated in terms of absolute joules by Ginnings and Furukawa.[13] Table 10.1 gives the results of the measurements of Osborne, Stimson, and Ginnings[8] on the heat capacity of water, converted to absolute joules deg^{-1} g^{-1} at a pressure of 1 atm. Values from this table should be used to convert older calorimetric data obtained in terms of the heat capacity of water at a given temperature.

D. JOULE VS. CALORIE*

Notwithstanding the fact that practically all accurate calorimetric measurements made after about 1910 were actually based on electrical energy, most investigators continued until about 1930 to express their final results in such a way as to make it appear that the unit of energy was in some way still connected with the heat capacity of water. Actually, what they did was to convert their values, determined in international joules, into one or more of the several calories based on the heat capacity of water, usually for comparison with older values reported in calories in the literature. This procedure should have been reversed; that is, the older data should have been converted to the modern unit of energy. However, the conversion to the older unit, the calorie, was favored because most chemists and physicists were reluctant to change from their habits of thinking of energy in terms of a unit of the size of a calorie.

An important effort to accustom chemists and physicists to the use of the joule as the unit of energy was made by Washburn in connection with many (but not all) of the tabulations of chemical thermodynamic data in the *International Critical Tables,*[14] of which he was editor-in-chief. This attempt to change over to the joule was not popular. It appeared then that the calorie would at least have to be retained as the name of the unit of heat energy. It was also realized that there would have to be separated from the new calorie every association with the heat capacity of water, else all the thermodynamic values would have to be changed every time someone determined the heat capacity of water with an accuracy greater

*The material in Sections 10.D through 10.F is taken from Rossini, F. D., *Chemical Thermodynamics,* John Wiley and Sons, New York, 1950. With permission.

TABLE 10.1

The Heat Capacity of Water in the Range 0 to 100° C*

Temperature °C	Heat capacity at 1 atm abs. joules $deg^{-1} g^{-1}$	Temperature °C	Heat capacity at 1 atm abs. joules $deg^{-1} g^{-1}$
0	4.2176	55	4.1823
5	4.2021	60	4.1843
10	4.1921	65	4.1867
15	4.1857	70	4.1895
20	4.1818	75	4.1927
25	4.1795	80	4.1963
30	4.1784	85	4.2004
35	4.1781	90	4.2050
40	4.1785	95	4.2102
45	4.1794	100	4.2159
50	4.1806		

*Converted from international joules to absolute joules using the 1947 value of the National Bureau of Standards: 1 international J = 1.000165 ± 0.000025 absolute J (see text).

From Osborne, N. S., Stimson, H. F., and Ginnings, D. C., *J. Res. Nat. Bur. Stand.*, 23, 197, 1939.

than that already existing. It would also be necessary for the new calorie to have a size approximately equal to that of the traditional calorie.

The obvious solution was to have an artificial, conventional calorie, defined as equal to a given number of electrical joules, the unit in which the calorimetric measurements are actually made. The investigators would then report their results in terms of the unit in which the measurements are actually made, and, for the benefit of those who prefer to continue thinking of energy in terms of a unit having the name and size of the calorie, would also give the values in terms of the artificial calorie by using the conventional factor for the conversion.

In line with the foregoing development, there came into use independently about 1930 two different artificial, conventional, defined, "dry" calories, one in the engineering steam tables and the other in thermochemistry and chemical thermodynamics.

E. THE "DRY" STEAM CALORIE

The artificial, conventional calorie used in the engineering steam tables is designated as the I.T. calorie (cal_{IT} – International Table calorie, which was first defined in 1929 by the International Steam Table Conference[15] by the relation

$$1\ cal_{IT} = \frac{1}{860}\ \text{international watt-hour} = 4.18605\ \text{international joules:} \quad (1)$$

With the 1947 National Bureau of Standards relation between the international and the absolute watt, the same steam calorie would be equal to

$$1\ cal_{IT} = 4.18674\ \text{absolute joules.} \quad (2)$$

Because of slight differences in the relations between the international and absolute watts at the several national standardizing laboratories, by international agreement the steam calorie is rounded off to become:

$$1\ cal_{IT} = 4.1868\ \text{absolute joules.} \quad (3)$$

As a matter of historical interest, it may be mentioned that the steam calorie was defined so that its value would be near the value of the mean (0 to 100°C) calorie. As indicated by its definition, the steam calorie is independent of the heat capacity of water.

By common consent, the British thermal unit (Btu) used in the engineering steam tables is defined in terms of the steam calorie so as to retain the convenient relation

$$1\ cal_{IT}/\text{gram} = 1.8\ \text{Btu/lb.} \quad (4)$$

F. THE "DRY" THERMOCHEMICAL CALORIE

The artificial, conventional, "dry" calorie that was used after about 1930 in all the research laboratories in the United States dealing with thermochemistry and chemical thermodynamics was defined completely by the relation:[16-19]

1 cal = 4.1833 international joules. (5)

Beginning January 1, 1948, this calorie was redefined in terms of absolute joules, using the 1947 National Bureau of Standards relation between the international and absolute electrical units,[12] so that

1 cal = 4.184 (exactly) absolute joules. (6)

With this redefinition, the thermochemical calorie represents exactly the same quantity of energy as before, and all the values previously reported in terms of the thermochemical calorie remain unchanged. As is obvious from the definition, the thermochemical calorie is independent of the heat capacity of water.

The number 4.1833, which originally defined the thermochemical calorie in terms of international joules, now has no particular significance, though for historical interest it may be mentioned that it arose from the quotient 4.185/1.0004, through the attempt to hold to the factor 4.185, selected by the *International Critical Tables*[14] for the relation between the absolute joule and the 15° calorie, and the factor 1.0004, selected in 1930 as the then best ratio of the size of the international joule to the absolute joule.[10]

G. DISCUSSION

As has been made evident above, several points need to be kept in mind:

1. Calorimetric measurements of energy before about 1908 to 1910 were usually in terms of the heat capacity of water at a given temperature, the "wet" calorie.

2. Calorimetric measurements with electrical energy after about 1910 to 1948 were made in terms of international joules fixed by the 1908 international electrical standards.

3. Care must be taken to scrutinize carefully calorimetric investigations reported in the literature in the period from about 1910 to about 1930 to ascertain whether they were based on electrical energy or on the "wet" calorie, and which one in fact.

4. Calorimetric measurements with electrical energy made after January 1, 1948 are expected to have been made in terms of absolute joules, though careful scrutiny of the publications soon after 1948 should be made.

5. Calorimetric data of any age made in terms of the actual heat capacity of water at a given temperature, "wet" calories, may be converted to absolute joules with the values of Osborne, Stimson, and Ginnings given in Table 10.1 here.[8]

In this chapter most of the material relating to electrical standards is from Reference 1. Mueller and Rossini[20] give additional details regarding "wet" and "dry" calories. Rossini[21] gives a summary of the total problem.

It is expected that as time goes on most scientists and engineers may move to the use of joules and kilojoules as the unit of energy, in accordance with the International Metric System, as described in Chapter 3, and that calories and kilocalories will become historical units.

H. REFERENCES

1. *Electrical Units and Standards,* National Bureau of Standards Circular 60, U.S. Government Printing Office, Washington, D.C., 1920.
2. **Griffiths, E. H.,** *Phil. Trans. R. Soc.* (Lond.), 184, 361, 1893; *Proc. R. Soc.* (Lond.), 55, 23, 1894.
3. **Schuster, A. and Gannon, W.,** *Phil. Trans. R. Soc.* (Lond.), 186, 415, 1895.
4. **Callendar, H. L. and Barnes, H. T.,** *Proc. R. Soc.* (Lond.), 82, 390, 1909.
5. **Bousfield, W. R. and Bousfield, W. E.,** *Phil. Trans. R. Soc.* (Lond.), 211, 199, 1911.
6. **Jaeger, W. and von Steinwehr, H.,** *Sitz. Preuss. Akad. Wissenschaften,* 1915, 424, 1915.
7. **Laby, T. H. and Hercus, E. O.,** in *International Critical Tables,* Vol. 5, Washburn, E. W., Ed., McGraw-Hill, New York, 1929, 78.
8. **Osborne, N. S., Stimson, H. F., and Ginnings, D. C.,** *J. Res. Nat. Bur. Stand.,* 23, 197, 1939.
9. **Osborne, N. S., Stimson, H. F., and Fiock, E. F.,** *J. Res. Nat. Bur. Stand.,* 5, 411, 1930.

10. *Technical News Bulletin No. 156,* National Bureau of Standards, Washington, D.C., 1930, 29.
11. **Wensel, H. T.,** *J. Res. Nat. Bur. Stand.,* 22, 375, 1939.
12. National Bureau of Standards Circular C459, U.S. Government Printing Office, Washington, D.C., 1947.
13. **Ginnings, D. C. and Furukawa, G. T.,** *J. Am. Chem. Soc.,* 75, 522, 1953.
14. *International Critical Tables,* Vol. 5, Washburn, E. W., Ed., McGraw-Hill, New York, 1929.
15. Report of the International Steam Tables Conference (London, 1929)
16. **Rossini, F. D.,** *J. Res. Nat. Bur. Stand.,* 6, 1, 1931.
17. First Report of the Permanent Committee on Thermochemistry of the International Union of Chemistry, Paris, 1934.
18. **Rossini, F. D.,** *Chem. Rev.,* 18, 233, 1936.
19. **Rossini, F. D.,** *Chem. Rev.,* 27, 1, 1940.
20. **Mueller, E. F. and Rossini, F. D.,** *Am. J. Phys.,* 12, 1, 1944.
21. **Rossini, F. D.,** *Chemical Thermodynamics,* John Wiley and Sons, New York, 1950, chap. 5.

Chapter 11

THE FUNDAMENTAL PHYSICAL CONSTANTS

A. INTRODUCTORY COMMENTS

The progress of science throughout the world requires that there be an adequate system of communication among scientists. Scientists have to communicate a tremendous amount of information, much of which is quantitative in character. The transmission of quantitative information should be made without the introduction of extraneous errors arising from the use by different investigators of discordant values of the fundamental constants.

An experimental scientist makes measurements in his laboratory of certain physical phenomena, with instruments and apparatus calibrated in terms of the fundamental units of measurement – length, mass, and time. Most frequently, this investigator reports in the literature quantitative information which is not precisely alone the quantities he has measured, but these combined with certain fundamental constants to obtain quantities for comparison with related observations of other investigators. It is important, therefore, that the values of such fundamental constants be ones generally accepted by the scientific community.

For proper and convenient communication, without misunderstanding, different investigators must use the same values of the fundamental constants in reducing their respective data; otherwise, the reported quantities may be significantly different even though the quantities originally measured may actually have been in complete accord. It is important, therefore, for all of science, that we have continually available a currently acceptable self-consistent set of "best" values of the fundamental constants.

The need for fundamental constants was recognized in chemistry many years ago when there was established the chemists' international scale of atomic weights, so that chemists throughout the world might communicate their observations in a quantitative language readily understood by all scientists. In addition to a self-consistent set of values for the atomic weights, science and technology need also a self-consistent set of "best" values of the fundamental physical constants. This need is particularly strong in the communication of quantitative information on the physical, thermodynamic, spectral, and other properties of the chemical substances.*

Included among the fundamental physical constants for which reliable and, hopefully, internationally agreed upon values are needed by the scientific and technical community of the world are the following: the velocity of light, the Avogadro Constant, the Faraday Constant, the Planck Constant, the Gas Constant, the Boltzmann Constant, the pressure-volume product for 1 mol of a gas at 0°C and zero pressure, the molal volume of an ideal gas at 0°C and 1 atm pressure, the absolute temperature of the triple point of water, the charge on the electron, the First Radiation Constant, the Second Radiation Constant, the Einstein Constant relating mass and energy, the constant relating spectroscopic wave number and energy, the standard acceleration of gravity, the standard atmosphere, the Universal Gravity Constant, the atomic mass unit, the mole, the Spectroscopic Fine Structure Constant, the Rydberg Constant, and the Stefan-Boltzmann Constant.

B. HISTORICAL SUMMARY OF COMPILATIONS OF FUNDAMENTAL PHYSICAL CONSTANTS

The following is a historical summary of the work of compiling self-consistent sets of values of the fundamental physical constants over the past half century.[1]

In 1926 Volume 1 of the International Critical Tables appeared with a complete, self-consistent list of recommended values of the fundamental physical constants.[2]

In 1929 Birge, working substantially alone, showed that the then generally accepted value for the charge on the electron was in error because an incorrect value was used for the viscosity of air in making calculations arising from observations in the oil-drop experiments, and published a revised self-consistent set of the fundamental physical constants.[3]

In the period 1937 to 1939, reports on the fundamental physical constants were published by von Friesen,[4] Dunnington,[5] and Wensel.[6]

In 1941 and 1945 Birge returned to publication with further revised sets of the fundamental physical constants.[7,8]

In 1948 a report on these constants by Stille[9] appeared. Then, in 1951, Bearden and Watts[10] published an extensive recommended set of values for the fundamental physical constants.

In 1948 and 1951 Du Mond and Cohen[11,12] published a very extensive and thoroughly documented set of these constants.

*From Rossini, F. D., *Pure Appl. Chem.*, 9, 453, 1964. With permission.

The two self-consistent sets of the fundamental physical constants published in 1951, by Bearden and Watts[10] on the one hand and Du Mond and Cohen[12] on the other, were in substantial accord within their respective limits of uncertainty.

The set of recommended values from the 1951 publication of Du Mond and Cohen[12] served as the basis for a report of the National Research Council, United States, on the status of the value of the fundamental constants for physical chemistry as of July 1, 1951, which report was published in 1952.[13]

In 1954 Bearden, Earle, Minkowski, and Thomsen[14] published another complete set of values of these constants. This was followed in 1955 by an extended report from Cohen, Du Mond, Layton, and Rollett,[15] and in 1957 and 1959 by similar reports from Bearden and Thomsen.[16,17]

Meanwhile, other events, which had a very significant effect on the values of the fundamental physical constants, were taking place in the period 1955 to 1961, summarized as follows by the author:[1] *

In 1955, the Advisory Committee on Thermometry of the International Committee on Weights and Measures had obtained international approval for the definition of the absolute temperature of the triple point of water as exactly 273.16°K (to become effective January 1, 1960). Since the temperature of the conventional "ice point" is lower by 0.0100 ± 0.0001°C, the absolute temperature of the ice point thus became 273.1500 ± 0.0001°K. Prior to this time, the absolute temperature of the ice point had been obtained from experimental measurements, involving, directly or indirectly, the pressure-volume product of a gas at zero pressure, at the ice point (0°C) and at the steam point (100°C). In most of the sets of values of the fundamental constants referred to above, the value of the absolute temperature of the ice point had been selected as 273.160 ± 0.010°K. Thus, a very significant change in this constant was introduced, particularly for thermodynamic quantities, since the value of the gas constant R is obtained as the quotient of the pressure-volume product for one mole of gas at zero pressure and 0°C, divided by the absolute temperature of the ice point:

$$R = (PV)_{0^\circ C}^{P=0} / T_{0^\circ C} \qquad (1)$$

Then, in 1960 and 1961, came the unification of the chemists' and the physicists' scales of atomic weights, by action of the International Union of Pure and Applied Physics (IUPAP) and the International Union of Pure and Applied Chemistry (IUPAC). The former chemists' scale was based on one mole as 16 grams of naturally occurring oxygen, containing ^{16}O, ^{17}O and ^{18}O. The former physicists' scale was based on one mole as 16 grams of pure ^{16}O. The new scale of atomic weights is based on one mole as 12 grams of pure ^{12}C. To attain the new unified scale, the former chemists' values were decreased by 43 parts per million and the former physicists' values were decreased by 318 parts per million (see Chapter 7). For thermodynamicists interested in the value of gas constant R, the change in the absolute temperature of the ice point from 273.16 to 273.15 (a decrease of 37 parts per million in $T_{0^\circ C}$) and the change in the scale of atomic weights (a decrease of 43 parts per million in the size of one mole) almost cancelled one another (see Equation 1), leaving a net change of only −6 parts per million, a negligible amount.

The foregoing changes in the absolute temperature of the ice point and in the scale of atomic weights, along with the appearance of much new experimental data relating to the atomic constants, resulted in new adjustments of the values of the fundamental constants by Cohen and Du Mond[18] and by Bearden and Thomsen.[19] Early in 1963, the National Research Council (U.S.A.) Committee on Fundamental Constants approved a recommended set of values of the fundamental constants based on the latest work of Cohen and Du Mond, with suggestions by Bearden and Thomsen.

At the 1963 Conference of the International Union of Pure and Applied Chemistry in London, in early July, the foregoing recommended set of fundamental constants was brought to the attention of the Commission on Thermodynamics and Thermochemistry and the Commission on Physico-chemical Data and Standards. Both Commissions were in favor of having an internationally agreed-upon set of fundamental constants, and, at their suggestion, the Secretary-General of IUPAC requested the author, then Associate Member of the IUPAC Commission on Thermodynamics and Thermochemistry, and Chairman of its Task Group on Fundamental Constants, to represent IUPAC at the Second International Conference on Nuclidic Masses at Vienna, July 15–19, 1963, and to attend the meeting there of the Commission on Nuclidic Masses and Related Atomic Constants of IUPAP. At this conference, Cohen presented a paper entitled "1963 Status Report on the Fundamental Constants." At the same Conference, the author[21] presented a report entitled "The International Union of Pure and Applied Chemistry and its Interest in the Values of the Fundamental Constants," stating that IUPAC was anxious to reach accord with IUPAP on the use of a single, self-consistent set of fundamental constants, citing the adoption of the ^{12}C unified scale of atomic weights as an outstanding example of the fruits of co-operation, and indicating that the appropriate groups belonging to IUPAC had reviewed the Cohen-Du Mond set of constants and were prepared to recommend their use by chemists as "best" values, with the understanding that they would be subject to revision at appropriate intervals by the proper authorities. At the meeting of the IUPAP Commission on Nuclidic Masses and Related Atomic Constants during the same Conference, Cohen and Du Mond presented their report on "Recommended Values of the Physical Constants –

*See also Chapter 4, "The Basic Scale of Temperature," and Chapter 7, "The Scale of Atomic Weights," as appropriate.

1963."[22] The views of IUPAC on fundamental constants were also brought before this Commission by the author.

The IUPAP Commission on Nuclidic Masses and Related Atomic Constants then approved the following resolution for presentation to the General Assembly of the International Union of Pure and Applied Physics at Warsaw later in 1963:

"The Commission of Nuclidic Masses and Related Atomic Constants has encouraged two of its members, J. W. M. Du Mond and E. R. Cohen, to prepare a self-consistent list of the most probable values of the fundamental constants. This list was presented to the Second International Conference on Nuclidic Masses held in Vienna, July 15–19, 1963. The Commission expects that these values will be widely used and will help to remove many of the confusions that have arisen from the use of differing sets of constants. In addition, it is expected that the appearance of this list will encourage further experimental work aimed at improving our knowledge of these values."

This resolution was approved by the 1963 General Assembly of IUPAP.

The recommended values of the fundamental constants presented in the 1963 Cohen-Du Mond report, as received by the National Research Council (U.S.A.), by the IUPAP Commission on Nuclidic Masses and Related Atomic Constants, and by the IUPAC Commissions on Thermodynamics and Thermochemistry and Physicochemical Data and Standards, were given wide distribution in the various countries.*

The foregoing 1963 set of recommended values of the fundamental physical constants had essentially achieved international acceptance by the actions taken by the International Union of Pure and Applied Chemistry and the International Union of Pure and Applied Physics, as described above. It was then hoped that an appropriate formal international organization could be arranged to operate in the area of fundamental physical constants in much the same way as does the internationally recognized Commission on Atomic Weights, as explained in Chapters 7 and 8. However, the two tasks would have one significant difference in that the Commission on Atomic Weights deals with a large array of numerical values which, in detail, are essentially independent of one another, whereas a Commission on Fundamental Physical Constants would deal with a large array of numerical values, many of which are interrelated, requiring that the entire set of physical constants be made fully self consistent.

In 1969 Taylor, Parker, and Langenberg[23] published a complete, new, self-consistent, and fully documented set of fundamental physical constants, independent of any national or international organization. In this report the following points were emphasized: improved mathematical techniques had been worked out to yield a better value of the fine structure constant, α; determinations of e/h by means of the Josephson effect yielded values of enormously improved accuracy; new experimental data related to other physical constants had recently become available; new theoretical work had been completed. Taylor, Parker, and Langenberg[23] thus set the stage for some form of international agreement on a set of recommended values of the fundamental physical constants to replace the 1963 set.

In 1966 there was established under the International Council of Scientific Unions (ICSU) the Committee on Data for Science and Technology (CODATA), the purpose of which was to promote, encourage, and coordinate, on a worldwide basis, the collection, analysis, and compilation of numerical data for science and technology. For the proper coordination of such work on an international basis, there is needed an internationally accepted set of values of the fundamental physical constants.

In 1968 the author, as President of CODATA-ICSU for the term 1966–70, appointed, with the advice and approval of the Executive Board, a Task Group on Fundamental Constants with the mission of producing an internationally acceptable set of values of the fundamental physical constants. If was felt that, since ICSU was the organization to which adhered the principal scientific unions of the world, a set of constants produced under ICSU auspices by a group of the world's experts in the area of physical constants would automatically find acceptance by the scientific and technical community of the world. The membership of the CODATA Task Group on Fundamental Constants was as follows: E. R. Cohen (U.S.), Chairman; R. D. Deslattes (U.S.); H. E. Duckworth (Canada); A. Horsfield (U.K.); B. A. Mamyrin (U.S.S.R.); B. N. Oleinik (U.S.S.R.); H. Preston-Thomas (Canada); U. Stille (Germany); J. Terrien (France); and Y. Yamamoto (Japan). Among the foregoing were representatives of the national standardizing laboratories of the U.S., U.K., U.S.S.R., Germany, Canada, France, and Japan.

The CODATA Task Group on Fundamental Constants completed its initial mission in a period

*From Rossini, F. D., *Pure Appl. Chem.*, 9, 453, 1964. With permission.

of about 5 years, making its formal report at the 1973 Annual Meeting of CODATA in Stockholm. The results of that report are given in the next chapter and constitute the first really internationally approved set of recommended values of the fundamental physical constants.[24,51] It is hoped that CODATA will maintain its Task Group on Fundamental Constants on a continuing basis, so that, perhaps in another 10 years, say 1983, we may have a revised set of constants for the scientific and technical community of the world.

C. EXPERIMENTAL MEASUREMENT OF PHYSICAL CONSTANTS

In the past century improvements in the quality, design, control, and operation of experimental apparatus, and advances in the techniques and sophistication of measurement, have resulted in phenomenal increases in the precision and accuracy of the values assigned to the various physical constants. However advanced the measurement of a given physical constant, the basis of the determination may remain essentially unchanged. Following are some simple examples:

The velocity of light, c, involves a distance, ℓ and a time, t, according to the relation:

$$c = \ell/t. \tag{2}$$

The early measurements of the velocity of light by Michelson and others did indeed involve measurement of time over a specified distance on the surface of the earth.[50] Measurement of the velocity of light was revolutionized by the work of Evenson, Wells, Peterson, Danielson, and Day, who determined, by direct counting, the near infrared frequency of the methane stabilized helium-neon laser oscillating at 3.39 μm, with the wavelength being measured in terms of the Krypton-86 standard.[25,26] The new measurements reduce the uncertainty in the value of the velocity of light to about 0.004 ppm.[24]

The Faraday Constant, $\mathcal{F}$, is the quantity of electricity associated with 1 mol of electrons:

$$\mathcal{F} = Ne \tag{3}$$

where e is the charge on one electron. The determination of the Faraday Constant requires simply the measurement of the ratio of the quantity of electricity to the number of moles of electrons transporting that given quantity of electricity:

$$\mathcal{F} = (It)/(mz/M) = ItM/mz \tag{4}$$

Here I is the current, t is the time, m is the mass of substance transported, M is the gram-molecular weight of the substance transported, and z is the number of electrons per gram-molecular weight of the substance. Early measurements of the Faraday Constant were made by Rosa and Vinal[27] with the silver coulometer, by Washburn and Bates[28] with the iodine coulometer, and Vinal and Bates[44] with a comparison of the silver and iodine coulometers. The recent measurements, made at the National Bureau of Standards, include the following: by Craig, Hoffman, Law, and Hamer,[29] involving silver-perchloric acid measurements; by Marenko and Taylor,[30] involving measurements with benzoic and oxalic acids; and by Bower,[31] involving measurements with the iodine coulometer. In 1949 Hipple, Sommer, and Thomas[45] published a measurement of the Faraday using an electromagnetic method in which the essential observation is the ratio of the charge to the mass of the hydrogen ion, H^+.

It is interesting to note the historical trend of values of the Faraday as determined with the silver coulometer, expressed as the electrochemical equivalent of silver, in milligrams per coulomb, as given by Horsfield:[42] 1884, Rayleigh and Sidgwick, 1.11794; 1884, Kohlrausch and Kohlrausch, 1.1183; 1904, Guthe, 1.11773; 1908, Smith, Mather, and Lowry, 1.11827 ± 0.00002; 1960, Craig, Hoffman, Law, and Hamer, 1.117972 ± 0.000019. In 1968 Hamer[43] reviewed the then existing data and, correcting to absolute electrical units and the new scale of atomic weights, obtained the following values for the Faraday, in coulombs per equivalent: 1908, silver (deposition), 96478; 1914, iodine (oxidation), 96489 ± 3; 1953, oxalate (oxidation), 96482.8 ± 3.0; 1953, electromagnetic, 96522 ± 3; 1960, silver (dissolution), 96486.6 ± 1.6.

One path to the evaluation of the Avogadro Constant, N, is the following: given is a pure, perfect crystalline substance of molecular weight, M, with precise measurement made of its density, d. Then with careful and precise X-ray measurements made on the crystalline substance at a given temperature and pressure, one can calculate the volume, v, assignable to one molecule of the material. The total volume, V, of the given sample of mass, m, is related to the mass and density:

$$V = m/d. \tag{5}$$

The Avogadro Constant is the number of molecules per mole, which in the given sample is

$$N = (V/v)/(m/M). \tag{6}$$

This becomes

$$N = (m/dv)/(m/M) = M/dv. \tag{7}$$

Thus, the evaluation of the Avogadro Constant requires a pure, perfect crystalline substance, measurement of its density, and determination of the crystal spacings from X-ray measurements. One of the big problems in such determinations has been the calibration of the X-ray unit of length in terms of the absolute unit of length. According to Cohen and Taylor,[24] the uncertainty in this relation has been reduced to about 2 ppm.

Values of the acceleration of gravity, g, have been obtained through a network of 25,000 absolute gravimeter and pendulum measurements developed under the auspices of the International Union of Geodesy and Geophysics and known as the International Gravity Standardization Net of 1971. The value of g is a combination of the attraction of the earth and rotational centripetal acceleration. Variations in the value of g are due to several effects: oblateness of the earth's figure; variation in the centrifugal force with latitude; variation in altitude on the surface of the earth; fluctuations in the density of the earth's crust; gravitational effects due to extraterrestrial bodies. The value of g changes by about 30 ppm for each 100-m change in altitude on the earth's surface. According to Cohen and Taylor,[24] the present values of g are good to about 0.02 ppm.

The value of the pressure-volume product for 1 mol of a gas at zero pressure and 0°C is one of the least reliable fundamental physical constants. Its value is determined, as illustrated in Figure 4.1 of Chapter 4, by measuring, at a series of reduced pressures toward zero pressure, the pressure and volume on a given mass, m, of a pure gas of molecular weight, M. These measurements yield the value of

$$(PV)^{P=0}_{0^\circ C} = (PV)^{ideal}_{0^\circ C} \tag{8}$$

which permits calculation of the molal volume of the ideal gas at 0°C and one atm, and of the value of the gas constant, R, in accordance with Equation 1.

In 1941 Cragoe[32] published an extensive review of the existing data leading to the evaluation of (PV) at zero pressure and 0°C, and, by utilizing data at high pressures with sophisticated extrapolation to zero pressure, obtained a value which has been generally accepted since then. Recently Batuecas[33] reviewed the existing data and changed Cragoe's value slightly. Cohen and Taylor[24] adjusted Batuecas' value since the latter had used the old definition of the liter (see Chapter 2). According to Cohen and Taylor, the new value has a reliability of 31 ppm.

The value of the Newtonian Gravitational Constant, G is the least reliable of our fundamental physical constants. Determination of the value of G involves measurements of the force, F, between two known masses at a known distance apart in accordance with the relation,

$$F = G\, m_1 m_2 / \ell^2 \tag{9}$$

where m_1 and m_2 are the masses of the bodies and ℓ is the distance between them. It is interesting to note the historical record or the determinations of the value of G, as reported by Beams,[34] the values being expressed in terms of units of 10^{-11} Nm^2 kg^{-2}, 10^{-11} m^3 s^{-2} kg^{-1}, or 10^{-8}dyn·cm^2·g^{-2}: 1798, Cavendish, 6.754; 1881, von Jolly, 6.465; 1889, Wilsing, 6.596; 1894, Poynting, 6.698; 1895, Boys, 6.658; 1896, Braun, 6.658; von Eötvös, 6.65; 1909, Cremieu, 6.67; 1930, Heyl, 6.670; 1942, Heyl and Chrzanowski, 6.673. Cohen and Taylor[24] have reviewed in detail the work of Heyl,[35] Heyl and Chrzanowski,[36] Rose, Parker, Lowry, Kuhlthau, and Beams,[37] and Pontikis[38] and conclude that the best value of G as of 1973 has an uncertainty of 615 ppm.

In former years the value of the Planck Constant, h, was determined from various spectroscopic, X-ray, and electron measurements, coupled with a number of different relations involving the velocity of light, c, the charge of the electron, e, the mass of the electron, m_e, the Faraday Constant, F, and the Avogadro Constant, N, as follows: measurements of the quantum limit of the continuous X-ray spectrum, to give h/e; electron diffraction measurements of DeBroglie wavelengths for electrons accelerated with a measured voltage, to give $h/(em_e)^{1/2}$; X-ray photoelectrons ejected with known quantum energies, and measured by magnetic deflection, to give $e^2/m_e h$; determination of the spectroscopic fine structure constant, to give e^2/hc; spectroscopic measurements of the Rydberg Constant to give $m_e e^4/h^3 c$. These measurements have been well reviewed by Du Mond and Cohen.[11]

In 1962 Josephson[39] predicted that if two weakly coupled superconductors are maintained at a direct current potential difference V, there exists an alternating current between the superconductors having a frequency, ν, such that

$$\nu = 2\, Ve/h \qquad (10)$$

where e is the charge on the electron. Figure 2.2 of Chapter 2 is a simple diagram illustrating the Josephson junction. The existence of this effect was confirmed by a number of investigators, as listed in a number of review articles by Josephson,[40] Taylor, Parker, and Langenberg,[23] Langenberg,[41] and Cohen and Taylor.[24] It is seen that the determination of the value of e/h by this method reduces to a measurement of a frequency and a corresponding direct current voltage:

$$e/h = \nu/2V. \qquad (11)$$

According to Cohen and Taylor,[24] values of e/h obtained by this method are reliable to near 0.1 ppm.

The Stefan-Boltzmann Constant, σ, is one which can be determined experimentally and also by derivation from a combination of other constants to which it is equivalent:

$$\sigma = 2\,\pi^5 k/15h^3c^2. \qquad (12)$$

Here k is the Boltzmann Constant, h is the Planck Constant, and c is the velocity of light. Cohen and Taylor find that the value of σ determined via Equation 12 has an uncertainty of about 125 ppm, whereas the recent and most accurate determination of the value of σ experimentally, by Blevin and Brown,[46] has an uncertainty of the order of 500 ppm.

One of the most important experimentally determined constants is the Rydberg Constant, R, which is related to other constants as follows:[47]

$$R_\infty = m_e\, e^4\, 2\,\pi^2/h^3 c \text{ (in cgs electrostatic units)} \qquad (13)$$

or

$$R_\infty = (\mu_o)^2\, m_e c^3 e^4/8h^3 \text{ (in SI units)} \qquad (14)$$

By this definition, the units of $R\infty$ are of reciprocal length. In these equations, m_e is the mass of the electron, μ_o is the permeability of free space ($4\pi \times 10^{-7}$ henry/meter). Series[47] gives a review of the measurements of the Rydberg Constant to 1970 and Cohen and Taylor[24] give a summary of the modern measurements (1968 to 1972). According to Cohen and Taylor,[24] the experimentally determined value of the Rydberg Constant is reliable to better than 0.1 ppm, and thus serves as an important input to the determination of the combination, $m_e e^4/h^3 c$.

D. CLASSIFICATION OF THE FUNDAMENTAL PHYSICAL CONSTANTS

The complete list of values of 100 or more different physical constants constitutes an imposing and formidable array of numbers. For the expert in fundamental constants the list is easily comprehended. For the average scientist, whose expertise lies elsewhere but who needs to use the internationally accepted "best" values in his work, a simpler presentation can be most helpful to his understanding of this important bulwark of science and technology.

Years ago, for his own better understanding of the problem, the author devised a simple classification of the physical constants.[49] This is an arbitrary classification of fundamental physical constants into three categories:

1. Those "defined" constants for which the values are fixed simply by definition by an appropriate international body;
2. Those "basic" constants for which the values are determined by experimental measurement;
3. Those "derived" constants for which the values are obtained from physical relations in-

volving values of the "basic" and "defined" constants.

Illustrative of this scheme of classification are the values of the 1963 fundamental constants for physical chemistry as presented by Rossini[1] in 1964, shown in Table 11.1. While this is in principle a simple and logical scheme of classification, conducive to good understanding on the part of those not expert in the area of physical constants, it is important to note that the selection of certain constants as "basic" ones and other constants as "derived" ones is arbitrary.[1] A least squares adjustment, such as that carried out by Cohen and Du Mond,[11,12] treats on an equal basis all constants which are determinable experimentally with required uncertainty. The simple classification presented in Table 11.1 simply takes over after the least squares adjustments of the entire calculation matrix are completed and the resulting values obtained.

In their new report Cohen and Taylor[24] classify the input data used in their least squares adjustment into two groups. The first group, labeled as the auxiliary constants, contains the experimentally determined quantities which have uncertainties sufficiently small that they can be considered as insignificant uncertainties for the present purposes. The second group contains the remaining experimentally determined quantities, excluding those whose uncertainties far exceed the uncertainties obtainable by derivation from the values of other constants in the first or second groups.

The first group of "more precise" quantities, as identified by Cohen and Taylor,[24] includes the following: $2e/h$, via the Josephson Effect; the velocity of light, c; acceleration of gravity, g; magnetic moment of the proton in terms of the Bohr magneton, μ_p/μ_B; atomic masses and mass ratios; Rydberg Constant for infinite mass, R_∞.

The second group of "less precise" quantities, as given by Cohen and Taylor,[24] include the following: Faraday Constant, F; proton gyromagnetic ratio, γ_p; X-ray length conversion factor, Λ; Avogadro Constant, via X-ray measurements and density; Compton electron wavelength, $h/m_e e$; fine structure constant.

Cohen and Taylor[24] also give values for three other quantities that are of "lesser precision": Newtonian Gravitational Constant G; the pressure-volume product of a gas at zero pressure and 0°C; and the Stefan-Boltzmann Constant, σ. The uncertainties in these three quantities, as given by Cohen and Taylor, are, in parts per million, respectively, 615, 31, and 125.

TABLE 11.1

Simple Classification of the Fundamental Constants for Physical Chemistry

"Defined" constants	
Unified atomic mass unit	u
Mole	mol
Standard acceleration of gravity, in free fall	g
Normal atmosphere, pressure	atm
Absolute temperature of the triple point of water	T_{tp}
Thermochemical calorie	cal
International steam calorie	cal_{IT}
"Basic" constants	
Velocity of light *in vacuo*	c
Avogadro number	N
Faraday constant	F
Planck constant	h
Pressure-volume product for 1 mol of gas at 0°C and zero pressure	$(PV)^{P=0}_{0°C}$
"Derived" constants	
Elementary charge	$e = F/N$
Gas constant	$R = (PV)^{P=0}_{0°C} / T_{0°C}$
Boltzmann constant	$k = R/N$
Second radiation constant	$c_2 = hc/k$
Einstein constant relating mass and energy	$Y = c^2$
Constant relating wave number and energy	$Z = Nhc$

From Rossini, F. D., *Pure Appl. Chem.*, 9, 453, 1964. With permission.

E. DISCUSSION

A substantially complete picture of the determination and evaluation of the values of the fundamental physical constants may be obtained from the following three documents: Taylor, Parker, and Langenberg,[23] Langenberg and Taylor,[48] and Cohen and Taylor.[24]

Significant improvements in the reliability of a number of the physical constants, as given in the recommended values of 1973 and as compared with those of 1963, are summarized by Cohen and

Taylor[24] as follows, in parts per million: charge on the electron, 12 to 2.9; Planck Constant, 24 to 5.4; mass of the electron, 14 to 5.1; Avogadro Constant, 15 to 5.0; Faraday Constant, 5.2 to 2.8.

Discussions of the values of the physical constants, from the standpoint of investigators expert in other fields who need reliable and internationally accepted values of the fundamental constants in their work, are given by Rossini et al.[1,13,49]

F. REFERENCES

1. **Rossini, F. D.,** *Pure Appl. Chem.,* 9, 453, 1964.
2. *International Critical Tables,* Vol. 1, Washburn, E. W., Ed., McGraw-Hill, New York, 1926.
3. **Birge, R. T.,** *Rev. Mod. Phys.,* 1, 1, 1929.
4. **von Friesen, S.,** *Proc. R. Soc. Ser. A,* 160, 424, 1937.
5. **Dunnington, F. G.,** *Rev. Mod. Phys.,* 11, 68, 1939.
6. **Wensel, H. T.,** *J. Res. Nat. Bur. Stand.,* 22, 375, 1939.
7. **Birge, R. T.,** *Rev. Mod. Phys.,* 13, 233, 1941.
8. **Birge, R. T.,** *Am. J. Phys.,* 13, 63, 1945.
9. **Stille, U.,** *Z. Phys.,* 125, 174, 1948.
10. **Bearden, J. A. and Watts, H. M.,** *Phys. Rev.,* 81, 73, 1951.
11. **Du Mond, J. W. M. and Cohen, E. R.,** *Rev. Mod. Phys.,* 20, 82, 1948.
12. **Du Mond, J. W. M. and Cohen, E. R.,** *Phys. Rev.,* 82, 555, 1951.
13. **Rossini, F. D., Gucker, F. T., Jr., Johnston, H. L., Pauling, L., and Vinal, G. W.,** *J. Am. Chem. Soc.,* 74, 2699, 1952.
14. **Bearden, J. A., Earle, M. D., Minkowski, J. M., and Thomsen, J. S.,** *Phys. Rev.,* 93, 629, 1954.
15. **Cohen, E. R., Du Mond, J. W. M., Layton, T. W., and Rollett, J. S.,** *Rev. Mod. Phys.,* 27, 363, 1955.
16. **Bearden, J. A. and Thomsen, J. S.,** *Nuovo Cimento, Suppl.,* 5, 267, 1957; 6, 141, 1957.
17. **Bearden, J. A. and Thomsen, J. S.,** *Am. J. Phys.,* 27, 569, 1959.
18. **Cohen, E. R. and Du Mond, J. W. M.,** Report to the American Physics Society, Washington, D.C., April 1962.
19. **Bearden, J. A. and Thomsen, J. S.,** Report to the American Physics Society, Washington, D.C., April 1962.
20. **Cohen, E. R.,** Second International Conference on Nuclidic Masses and Related Constants, International Union of Pure and Applied Physics, Vienna, July 1963.
21. **Rossini, F. D.,** Proceedings of the Second International Conference on Nuclidic Masses, Vienna, July 15–19, 1963, 99.
22. **Cohen, E. R. and Du Mond, J. W. M.,** *Rev. Mod. Phys.,* 37, 537, 1965.
23. **Taylor, B. N., Parker, W. H., and Langenberg, D. N.,** *The Fundamental Constants and Quantum Electrodynamics: A Review of Modern Physics Monograph,* Academic Press, New York, 1969.
24. **Cohen, E. R. and Taylor, B. N.,** *J. Phys. Chem. Ref. Data,* 2, 663, 1973.
25. **Evenson, K. M., Wells, J. S., Peterson, F. R., Danielson, B. L., and Day, G. W.,** *Appl. Phys. Lett.,* 22, 192, 1973.
26. **Barger, R. L. and Hall, J. L.,** *Appl. Phys. Lett.,* 22, 196, 1973.
27. **Rosa, E. B. and Vinal, G. W.,** *Bull. Bur. Stand.,* 9, 151, 1913.
28. **Washburn, E. W. and Bates, S. J.,** *J. Am. Chem. Soc.,* 34, 1341, 1515, 1912.
29. **Craig, D. N., Hoffman, J. I., Law, C. A., and Hamer, W. J.,** *J. Res. Nat. Bur. Stand.,* 64A, 381, 1960.
30. **Marenko, G. and Taylor, J. K.,** *Anal. Chem.,* 40, 1645, 1968.
31. **Bower, V. E.,** in *Atomic Masses and Fundamental Constants,* Sanders, J. H. and Wapstra, A. H., Eds., Plenum, New York, 1972, 516.
32. **Cragoe, C. S.,** *J. Res. Nat. Bur. Stand.,* 26, 495, 1941.
33. **Batuecas, T.,** in *On Atomic Masses and Fundamental Constants,* Sanders, J. H. and Wapstra, A. H., Eds., Plenum, New York, 1972, 534.
34. **Beams, J. W.,** *Phys. Today,* 24, 35, 1971.
35. **Heyl, P. R.,** *J. Res. Nat. Bur. Stand.,* 5, 1243, 1930.
36. **Heyl, P. R. and Chrzanowski, P.,** *J. Res. Nat. Bur. Stand.,* 29, 1, 1942.
37. **Rose, R. D., Parker, H. M., Lowry, R. A., Kuhlthau, A. R., and Beams, J. W.,** *Phys. Rev. Lett.,* 23, 655, 1969.
38. **Pontikis, C.,** *C. R. Hebd. Seances Acad. Sci. Ser. B,* 274, 437, 1972.
39. **Josephson, B. D.,** *Phys. Lett.,* 1, 251, 1962.
40. **Josephson, B. D.,** *Advan. Phys.,* 14, 419, 1965.

41. **Langenberg, D. N.**, in *Precision Measurement and Fundamental Constants,* Langenberg, D. N. and Taylor, B. N., Eds., National Bureau of Standards Special Publication 343, U.S. Government Printing Office, Washington, D.C., 1971, 203.
42. **Horsfield, A.**, in *Precision Measurement and Fundamental Constants,* Langenberg, D. N. and Taylor, B. N., Eds., National Bureau of Standards Special Publication 343, U.S. Government Printing Office, Washington, D.C., 1971, 137.
43. **Hamer, W. J.**, *J. Res. Nat. Bur. Stand.*, 72A, 435, 1968.
44. **Vinal, G. W. and Bates, S. J.**, *Bull. Bur. Stand.*, 10, 425, 1914.
45. **Hipple, J. A., Sommer, H., and Thomas, H. A.**, *Phys. Rev.*, 76, 1877, 1949.
46. **Blevin, W. R. and Brown, W. J.**, *Metrologia,* 7, 15, 1971.
47. **Series, G. W.**, in *Precision Measurement and Fundamental Constants,* Langenberg, D. N. and Taylor, B. N., Eds., National Bureau of Standards Special Publication 343, U.S. Government Printing Office, Washington, D.C., 1971, 73.
48. **Langenberg, D. N. and Taylor, B. N., Eds.**, *Precision Measurement and Fundamental Constants,* National Bureau of Standards Special Publication 343, U.S. Government Printing Office, Washington, D.C., 1971.
49. **Rossini, F. D.**, *Chemical Thermodynamics,* John Wiley and Sons, New York, 1950, chap. 4.
50. **Shankland, R. S.**, *Phys. Today,* 27, 37, 1974.
51. Report of the CODATA Task Group on Fundamental Constants, Cohen, E. R., chairman, CODATA Bulletin No. 11, December 1973, CODATA Central Office, Westendstrasse 19, 6 Frankfurt am Main, Germany (B.R.D.).

Chapter 12

THE INTERNATIONAL PHYSICAL CONSTANTS OF 1973

A. INTRODUCTORY COMMENTS

As explained in the preceding chapter, the International Physical Constants of 1973, which replaced the so-called international set of 1963, were produced by the Task Group on Fundamental Constants of CODATA, the Committee on Data for Science and Technology, under ICSU, the International Council of Scientific Unions. This CODATA Task Group consisted of ten of the world's experts on fundamental constants, representing seven countries and their national standardizing laboratories, as well as providing close liaison with other international groups interested in physical constants. Because of the expertise of the members of the Task Group, and the truly international and multidiscipline nature of its sponsor, ICSU, their 1973 set of recommended values of the fundamental physical constants has become a truly international set, with full acceptance by the scientific and technical community of the world in much the same way as the international table of atomic weights is accepted.

In this chapter we present the 1973 international values of the physical constants in the form presented by the CODATA Task Group, as published in the report of Cohen and Taylor,[1,3] together with other arrangements which may be more helpful to those not expert in physical constants. Also presented are tables of conversion factors to permit easy conversion from old to modern units.

B. THE 1973 INTERNATIONAL PHYSICAL CONSTANTS

Table 12.1 gives the values recommended for the 1973 set of International Physical Constants, from Cohen and Taylor,[1,3] as recommended to CODATA and ICSU by the CODATA Task Group on Fundamental Constants. As indicated by the headings of Table 12.1, the columns give, in order, the quantity, symbol, value (with the standard deviation in units of the last figure given in parentheses), the uncertainty in parts per million (measured by the standard deviation), the units in the international metric system (SI), and the units in the cgs system.

Table 12.2 gives a collection of values of fundamental physical constants which the author identifies as "defined," as explained in the preceding chapter. In this connection, the following definitions should be noted:

1. The unified atomic mass unit, for which the symbol is u, is 1/12 times the mass of an atom of Carbon-12, ^{12}C;
2. The mole, for which the symbol is mol, is the amount of a substance, of specified chemical formula, containing the same number of formula units (molecules, atoms, ions, electrons, or other entities) as there are atoms in 12 g (exactly) of the pure nuclide, Carbon-12, ^{12}C.

Also, it should be noted, as explained in Chapter 2, that the inch (U.S.) has been defined by the relation

$$1 \text{ in.} = 2.54 \text{ (exactly) cm} \quad (1)$$

and the pound (U.S.) has been defined by the relation

$$1 \text{ lb} = 453.59237 \text{ (exactly) g.} \quad (2)$$

Table 12.3 gives a collection of values of some fundamental physical constants, extracted from Table 12.1, which the author identifies as "basic," that is, evaluated from experimental measurements as explained in the preceding chapter.

Table 12.4 gives a collection of values of some fundamental physical constants, which the author identifies as "derived," which are obtained by using appropriate physical relations with values from Tables 12.1, 12.2, or 12.3.

The values in Tables 12.3 and 12.4 are given in both cgs and SI units, and the uncertainty, labeled "Δ," is given in parts per million, ppm. As previously mentioned, the identification of some physical constants as "basic" and the others (excluding those which are defined) as "derived" is arbitrary and is done to simplify the understanding of those not expert in the area of physical constants.

TABLE 12.1

Values[a] of the 1973 International Physical Constants as Recommended by CODATA-ICSU

Quantity	Symbol	Value	Uncertainty (ppm)	Units	
				SI	cgs
Speed of light in vacuum	c	2.997924580(12)	0.004	10^{8} m·s^{-1}	10^{10} cm·s^{-1}
Fine-structure Constant, $[\mu_0 c^2/4\pi]$ $(e^2/\hbar c)$	α	7.2973506(60)	0.82	10^{-3}	10^{-3}
	α^{-1}	137.03604(11)	0.82		
Electron charge	e	1.6021892(46)	2.9	10^{-19} C	10^{-20} emu
		4.803242(14)	2.9		10^{-10} esu
Planck Constant	h	6.626176(36)	5.4	10^{-34} J·s	10^{-27} erg·s
	$\hbar = h/2\pi$	1.0545887(57)	5.4	10^{-34} J·s	10^{-27} erg·s
Avogadro Constant	N	6.022045(31)	5.1	10^{26} kmol^{-1}	10^{23} mol^{-1}
Atomic mass unit	u	1.6605655(86)	5.1	10^{-27} kg	10^{-24} g
Electron rest mass	m_e	9.109534(47)	5.1	10^{-31} kg	10^{-28} g
$Nm_e = M_e$	M_e	5.4858026(21)	0.38	10^{-4} u	10^{-4} u
Proton rest mass	m_p	1.6726485(86)	5.1	10^{-27} kg	10^{-24} g
$Nm_p = M_p$	M_p	1.007276471(11)	0.011	u	u
Ratio of proton mass to electron mass	m_p/m_e	1836.15152(70)	0.38		
Neutron rest mass	m_n	1.6749543(86)	5.1	10^{-27} kg	10^{-24} g
$Nm_n = M_n$	M_n	1.008665012(37)	0.037	u	u
Electron charge to mass ratio	e/m_e	1.7588047(49)	2.8	10^{11} C·kg^{-1}	10^{7} emu·g^{-1}
		5.272764(15)	2.8		10^{17} esu·g^{-1}
Magnetic flux quantum, $[c]^{-1}$ $(hc/2e)$	Φ_0	2.0678506(54)	2.6	10^{-15} Wb	10^{-7} G·cm^2
	h/e	4.135701(11)	2.6	10^{-15} J·s·C^{-1}	10^{-7} erg·s·emu^{-1}
		1.3795215(36)	2.6		10^{-17} erg·s·esu^{-1}
Josephson frequency-voltage ratio	$2e/h$	4.835939(13)	2.6	10^{14} Hz·V^{-1}	
Quantum of circulation	$h/2m_e$	3.6369455(60)	1.6	10^{-4} J·s·kg^{-1}	erg·s·g^{-1}
	h/m_e	7.273891(12)	1.6	10^{-4} J·s·kg^{-1}	erg·s·g^{-1}
Faraday Constant	$\mathcal{F}$	9.648456(27)	2.8	10^{7} C·kmol^{-1}	10^{4} C·mol^{-1}
		2.8925342(82)	2.8		10^{14} esu·mol^{-1}

[a]Note that the numbers in parentheses are the one standard deviation uncertainties in the last digits of the quoted value computed on the basis of international consistency, that the unified atomic mass scale $^{12}C \equiv 12$ has been used throughout, that u = atomic mass unit, c = coulomb, G = gauss, Hz = hertz = cycles/sec, J = joule, K = kelvin (degrees kelvin), T = tesla, (10^4 G), V = volt, Wb = weber = T·m^2, and W = watt. In the cases where formulas for constants are given (e.g., R_∞), the relations are written as the product of two factors. The second factor, in parentheses, is the expression to be used when all quantities are expressed in cgs units, with the electron charge in electrostatic units. The first factor, in brackets, is to be included only if all quantities are expressed in SI units. We remind the reader that with the exception of the auxiliary constants which have been taken to be exact, the uncertainties of these constants are correlated, and therefore the general law of error propagation must be used in calculating additional quantities requiring two or more of these constants.

TABLE 12.1 (continued)

Values[a] of the 1973 International Physical Constants as Recommended by CODATA-ICSU

Quantity	Symbol	Value	Uncertainty (ppm)	Units SI	Units cgs
Rydberg constant, $[\mu_0 c^2/4\pi]^2 (m_e e^4/4\pi\hbar^3 c)$	R_∞	1.097373177(83)	0.075	10^7 m^{-1}	10^5 cm^{-1}
Bohr radius, $[\mu_0 c^2/4\pi]^{-1} (\hbar^2/m_e e^2) = \alpha/4\pi R_\infty$	a_0	5.2917706(44)	0.82	10^{-11} m	10^{-9} cm
Classical electron radius, $[\mu_0 c^2/4\pi](e^2/m_e c^2) = \alpha^3/4\pi R_\infty$	r_0	2.8179380(70)	2.5	10^{-15} m	10^{-13} cm
Free electron g-factor, or electron magnetic moment in Bohr magnetons	$g_{j/2} = \mu_e/\mu_B$	1.0011596567(35)	0.0035		
Free muon g-factor, or muon magnetic moment in units of [c] $(e\hbar/2m_\mu c)$	$g_{\mu/2}$	1.00116616(31)	0.31		
Bohr magneton, [c] $(e\hbar/2m_e c)$	μ_B	9.274078(36)	3.9	10^{-24} $J{\cdot}T^{-1}$	10^{-21} $erg{\cdot}G^{-1}$
Electron magnetic moment	μ_e	9.284832(36)	3.9	10^{-24} $J{\cdot}T^{-1}$	10^{-21} $erg{\cdot}G^{-1}$
Gyromagnetic ratio of protons in H_2O	γ'_p	2.6751301(75)	2.8	10^8 $rad{\cdot}s^{-1}{\cdot}T^{-1}$	10^4 $rad{\cdot}s^{-1}{\cdot}G^{-1}$
	$\gamma'_p/2\pi$	4.257602(12)	2.8	10^7 $Hz{\cdot}T^{-1}$	10^3 $Hz{\cdot}G^{-1}$
γ'_p corrected for diamagnetism of H_2O	γ_p	2.6751987(75)	2.8	10^8 $rad{\cdot}s^{-1}{\cdot}T^{-1}$	10^4 $rad{\cdot}s^{-1}{\cdot}G^{-1}$
	$\gamma_p/2\pi$	4.257711(12)	2.8	10^7 $Hz{\cdot}T^{-1}$	10^3 $Hz{\cdot}G^{-1}$
Magnetic moment of protons in H_2O in Bohr magnetons	μ'_p/μ_B	1.52099322(10)	0.066	10^{-3}	10^{-3}
Proton magnetic moment in Bohr magnetons	μ_p/μ_B	1.521032209(16)	0.011	10^{-3}	10^{-3}
Ratio of electron and proton magnetic moments	μ_e/μ_p	658.2106880(66)	0.010		
Proton magnetic moment	μ_p	1.4106171(55)	3.9	10^{-26} $J{\cdot}T^{-1}$	10^{-23} $erg{\cdot}G^{-1}$
Magnetic moment of protons in H_2O in nuclear magnetons	μ'_p/μ_N	2.7927740(11)	0.38		
μ'_p/μ_N corrected for diamagnetism of H_2O	μ_p/μ_N	2.7928456(11)	0.38		
Nuclear magneton, [c] $(e\hbar/2m_p c)$	μ_N	5.050824(20)	3.9	10^{-27} $J{\cdot}T^{-1}$	10^{-24} $erg{\cdot}G^{-1}$
Ratio of muon and proton magnetic moments	μ_μ/μ_p	3.1833402(72)	2.3		
Muon magnetic moment	μ_μ	4.490474(18)	3.9	10^{-26} $J{\cdot}T^{-1}$	10^{-23} $erg{\cdot}G^{-1}$

TABLE 12.1 (continued)

Values[a] of the 1973 International Physical Constants as Recommended by CODATA-ICSU

Quantity	Symbol	Value	Uncertainty (ppm)	Units SI	Units cgs
Ratio of muon mass to electron mass	m_μ/m_e	206.76865(47)	2.3		
Muon rest mass	m_μ	1.883566(11)	5.6	10^{-28} kg	10^{-25} g
	M_μ	0.11342920(26)	2.3	u	u
Compton wavelength of the electron, h/m_ec	λ_C	2.4263089(40)	1.6	10^{-12} m	10^{-10} cm
	$\lambda_C/2\pi$	3.8615905(64)	1.6	10^{-13} m	10^{-11} cm
Compton wavelength of the proton h/m_pc	$\lambda_{C,p}$	1.3214099(22)	1.7	10^{-15} m	10^{-13} cm
	$\lambda_{C,p}/2\pi$	2.1030892(36)	1.7	10^{-16} m	10^{-14} cm
Compton wavelength of the neutron, h/m_nc	$\lambda_{C,n}$	1.3195909(22)	1.7	10^{-15} m	10^{-13} cm
	$\lambda_{C,n}/2\pi$	2.1001941(35)	1.7	10^{-16} m	10^{-14} cm
Standard volume of ideal gas	V_0	22.71081(71)	31	10^5 J·kmol^{-1}	10^9 erg·mol^{-1}
		22.41383(70)	31	m^3·atm·kmol^{-1}	10^3 cm^3·atm·mol^{-1}
Gas constant, V_0/T_0 (T_0 = 273.15 K)	R	8.31441(26)	31	10^3 J·kmol^{-1}·K^{-1}	10^7 erg·mol^{-1}·K^{-1}
		8.20568(26)	31	10^{-2} m^3·atm·kmol^{-1}·K^{-1}	10 cm^3·atm·mol^{-1}·K^{-1}
Boltzmann constant, R/N	k	1.380662(44)	32	10^{-23} J·K^{-1}	10^{-16} erg·K^{-1}
Stefan-Boltzmann constant, $\pi^2k^4/60\hbar^3c^2$	σ	5.67032(71)	125	10^{-8} W·m^{-2}·K^{-4}	10^{-5} erg·s^{-1}·cm^{-2}·K^{-4}
First radiation constant, $2\pi hc^2$	c_1	3.741832(20)	5.4	10^{-16} W·m^2	10^{-5} erg·cm^2·s^{-1}
Second radiation constant, hc/k	c_2	1.438786(45)	31	10^{-2} m·K	cm·K
Gravitational constant	G	6.6720(41)	615	10^{-11} N·m^2·kg^{-2}	10^{-8} dyn·cm^2·g^{-2}

From Cohen, E. R. and Taylor, B. N., *J. Phys. Chem. Ref. Data*, 2, 663, 1973. With permission. (Also published in Report of the CODATA Task Group on Fundamental Constants, Cohen, E. R., Chairman, CODATA Bulletin No. 11, December, 1973. CODATA Central Office, ICSU, 51 Blvd. de Montmorency, F-75016 Paris, France.)

TABLE 12.2

Values of Some "Defined" Physical Constants

Constant	Symbol	Value (exact, by definition)
Standard acceleration of gravity, in free fall	g	980.665 cm s^{-2}
Normal atmosphere, pressure	atm	1,013,250 dyn cm^{-2}
Absolute temperature of the triple point of water	T_{tp}	273.16 K
Thermochemical calorie	cal	4.184 J
International steam calorie	cal_{IT}	4.1868 J

TABLE 12.3

Values of Some "Basic" (Experimentally Evaluated) Physical Constants

Quantity	Symbol	Value	Δ (ppm)	Units: cgs	Units: SI
Speed of light in vacuo	c	2.99792458	0.004	10^{10} cm·s^{-1}	10^{8} m·s^{-1}
Faraday constant	$\mathfrak{F}$	9.648456	2.8	10^{4} C·mol^{-1}	10^{7} C·$Kmol^{-1}$
Electronic charge	e	1.6021892	2.9	10^{-20} emu	10^{-19} C
Planck constant	h	6.626176	5.4	10^{-27} erg·s	10^{-34} J·s
PV product – one mol gas, P = 0, at 0°C	$(PV)_{0°C}^{P=0}$	22.41383	31	10^{3} cm^{3}·atm·mol^{-1}	10^{-3} m^{3}·atm·mol^{-1}
Gravitational constant	G	6.6720	615	10^{-3} dyn·cm^{2}·g^{-2}	10^{-11} N·m^{2}·kg^{-2}
Electron rest mass	M_e	5.4858026	0.38	10^{-4} g·mol^{-1}	10^{-4} kg·$Kmol^{-1}$
Neutron rest mass	M_n	1.008665012	0.037	g·mol^{-1}	kg·$Kmol^{-1}$
Proton rest mass	M_p	1.007276471	0.011	g·mol^{-1}	kg·$Kmol^{-1}$

Extracted from Table 12.1.

TABLE 12.4

Values of Some "Derived" Physical Constants, Obtained by Using Appropriate Physical Relations and Values of the "Defined" and "Basic" (Experimentally Evaluated) Constants Given in Tables 12.2 and 12.3

Quantity	Symbol	Value	Δ (ppm)	Units cgs	Units SI
Avogadro Constant, $N = \mathfrak{F}/e$	N	6.022045	5.1	10^{23} mol^{-1}	10^{23} mol^{-1}
Gas Constant, $R = (Pv)^{P=0}_{0°C}/T_{0°C}$ [9 5]	R	8.31441	31	J·mol^{-1}·K^{-1}	J·mol^{-1}·K^{-1}
Boltzmann Constant, $k = R/N$	k	1.380662	32	10^{-23} J·K^{-1}·molecule^{-1}	10^{-23} J·K^{-1}·molecule^{-1}
Second radiation Constant, $c_2 = hc/k$	c_2	1.438786	31	cm·K	10^{-2} m·K
Einstein Constant, relating mass and energy, $Y = c^2$	Y	8.9875517	0.006	10^{13} J·g^{-1}	10^{16} J·kg^{-1}
Constant relating wave number and energy, $Z = Nhc$	Z	11.962657	1.7	J·cm·mol^{-1}	10^{-2} J·m·mol^{-1}
First Radiation Constant, $c_1 = 2\pi hc^2$	c_1	3.741832	5.4	10^{-12} J·cm^2·S^{-1}	10^{-16} W·m^2
Stefan-Boltzmann constant, $\sigma = 2\pi^5 k/15h^3c^2$	σ	5.67032	125	10^{-12} J·s^{-1}·cm^{-2}·K^{-4}	10^{-8} W·m^{-2}·K^{-4}
Compton electron wavelength, $\lambda_c = h/cm_e$	λ_c	2.4263089	1.6	10^{-10} cm	10^{-12} m
Compton proton wavelength, $\lambda_{c,p} = h/cm_p$	$\lambda_{c,p}$	1.3214099	1.7	10^{-13} cm	10^{-15} m
Compton neutron wavelength, $\lambda_{c,n} = h/cm_n$	$\lambda_{c,n}$	1.3195909	1.7	10^{-13} cm	10^{-15} m
Fine structure Constant, $\alpha = \mu_0 ce^2/2h$	α	7.2973506	0.82	10^{-3}	10^{-3}
Atomic mass unit, u = $1/N$	u	1.6605655	5.1	10^{-24} g	10^{-27} Kg
Electron charge to mass ratio	e/m_e	1.7588047	2.8	10^7 emu·g^{-1}	10^{11} C·kg^{-1}

Derived from Tables 12.1, 12.2, and 12.3.

Table 12.5 gives values for some frequently used physical constants with the energy expressed in calories.

C. SOME CONVERSION FACTORS

Table 12.6, from Rossini,[2] with the new values of the constants used, gives some common conversion factors for the following properties: length, area, volume, mass, density, force, pressure, energy, molecular energy, specific energy, and specific energy per degree.

D. DISCUSSION

The uncertainties associated with the new 1973 International Physical Constants have reached the point where such uncertainties are negligible for essentially all scientific work, except in the case of those experts in physical constants who are seeking to improve the existing values and further reduce the uncertainties. Except for the Gravitational Constant, which has an assigned uncertainty of 615 ppm, and the Stefan-Boltzmann Constant, which has an assigned uncertainty of 125 ppm, all other physical constants have an assigned uncertainty of 31 ppm or less, with most of them being less than about 5 ppm.

With the new organizational structure for producing recommended International Physical Constants, under the International Council of Scientific Unions, we can look forward to having a group of the world's experts in the field monitoring new data on physical constants and, at the appropriate time, producing a new recommended set of the physical constants. As reviewed in the previous chapter, we see that there was a generally accepted set produced in 1952; then came a 1963 set, and now we have a 1973 set. Hopefully, we can look forward to a new set in about a decade.

The author is indebted for special information to E. Richard Cohen, Chairman of the CODATA-ICSU Task Group on Fundamental Constants.

TABLE 12.5

Values of Some Frequently Used Physical Constants with Energy Expressed in Calories

Quantity	Symbol	Value	Δ (ppm)	Units
Faraday Constant	$\mathcal{F}$	23060.36	2.8	$cal \cdot V^{-1} \cdot mol^{-1}$
Gas Constant	R	1.98719	31	$cal \cdot K^{-1} \cdot mol^{-1}$
Einstein Constant relating mass and energy	Y	2.14807643	0.006	$10^{13} cal \cdot g^{-1}$
Constant relating wave number and energy	Z	2.859144	1.7	$cal \cdot cm \cdot mol^{-1}$

TABLE 12.6

Some Common Conversion Factors

Length

1 angstrom (Å) = 10 nm = 10^{-8} cm = 10^{-10} m.
1 micron = 1 μm = 10^{-4} cm = 10^{-6} m.
1 inch (in.) = 2.54 (exactly) cm = 0.0254 (exactly) m.
1 foot (ft) = 30.48 (exactly) cm = 0.3048 (exactly) m.

Area

1 cm^2 = 10^{-4} m^2.
1 $in.^2$ = 6.4516 (exactly) cm^2 = 6.4516 × 10^{-4} (exactly) m^2.
1 ft^2 = 929.0304 (exactly) cm^2 = 0.09290304 (exactly) m^2.

TABLE 12.6 (continued)

Some Common Conversion Factors

Volume

1 $cm^3 = 10^{-6}$ m^3.
1 $in.^3$ = 16.387064 (exactly) cm^3 = 1.6387064×10^{-5} (exactly) m^3.
1 ft^3 = 28316.846592 (exactly) cm^3 = 0.028316846592 (exactly) m^3.
1 liter (l) = 1,000 (exactly) cm^3 = 1×10^{-3} (exactly) m^3.
1 gallon (U.S.) = 231 (exactly) $in.^3$ = 3785.411784 (exactly) cm^3 = $3.785411784 \times 10^{-3}$ (exactly) m^3.

Mass

1 g = 10^{-3} kg.
1 pound (lb) = 453.59237 (exactly) g = 0.45359237 (exactly) kg.
1 metric ton = 1 tonne = 10^3 kg = 10^6 g.

Density

1 $g/cm^3 = 10^3$ kg/m^3.
1 $lb/in.^3$ = 27.679905 g/cm^3 = 27679.905 kg/m^3.
1 lb/ft^3 = 0.016018463 g/cm^3 = 16.018463 kg/m^3.
1 lb/gal = 0.11982642 g/cm^3 = 119.82642 kg/m^3.

Force

1 dyn = 10^{-5} newton (N).

Pressure

1 dyn/cm^2 = 0.1 N/m^2.
1 bar = 10^6 dyn/cm^2 = 10^5 N/m^2.
1 atm = 1,013,250 (exactly) dyn/cm^2 = 1.013250 bar = 1.013250×10^5 N/m^2.
1 torr = 0.001315789 atm = 0.001333223 bar = 133.3223 N/m^2.
1 atm = 14.6960 lb (wt)/$in.^2$.

Energy

1 J = 10^7 erg.
1 cal = 4.184 (exactly) J.
1 cal_{IT} = 4.1868 (exactly) J.
1 Btu_{IT} = 1055.056 J = 251.9958 cal_{IT}.
1 kWh = 3.6×10^6 (exactly) J = 8.604206×10^5 cal = 8.59845×10^5 cal_{IT} = 3.412141×10^3 Btu_{IT}.
1 l-atm = 101.3250 J = 24.21726 cal.

Molecular energy

1 J/mol = 0.2390057 cal/mol.
1 eV/molecule = 96,484.56 J/mol = 23,060.36 cal/mol = 8065.479 wave no. (cm^{-1}).
1 wave no. (cm^{-1}) = 11.962657 J/mol = 2.859144 cal/mol.

Specific energy

1 cal/g = 4.184 (exactly) J/g.
1 cal_{IT}/g = 4.1868 (exactly) J/g = 1.8 (exactly) Btu_{IT}/lb.
1 Btu_{IT}/lb = 2.326000 J/g = 0.5555556 cal_{IT}/g.

TABLE 12.6 (continued)

Some Common Conversion Factors

Specific energy per degree

1 cal/g K = 4.184 (exactly) J/g K.
1 cal_{IT}/g K = 4.1868 (exactly) J/g K.
1 Btu_{IT}/lb deg F = 1 (exactly) cal_{IT}/g K.

Revised from Rossini, F. D., *Chemical Thermodynamics,* John Wiley and Sons, New York, 1950. With permission.

E. REFERENCES

1. **Cohen, E. R. and Taylor, B. N.,** *J. Phys. Chem. Ref. Data,* 2, 663, 1973.
2. **Rossini, F. D.,** *Chemical Thermodynamics,* John Wiley and Sons, New York, 1950, appendix, tables 2–13.
3. Report of the CODATA Task Group on Fundamental Constants, Cohen, E. R., Chairman, CODATA Bulletin No. 11, December, 1973. CODATA Central Office, ICSU, 51 Blvd. de Montmorency, F-75016 Paris, France.

Chapter 13

NUMERICAL DATA FOR SCIENCE AND TECHNOLOGY*

A. INTRODUCTORY COMMENTS

The problem of numerical data for science and technology has occupied the attention of scientists for many years. About 150 years ago there were in the world about 50 scientific journals, 100 years ago about 500, 50 years ago about 5,000, and today the number is of the order of 50,000. The quantity of scientific information has been increasing at an enormous rate, said to be doubling every 8 to 10 years.

It is estimated that more than one million scientific and technical papers are appearing each year in journals, bulletins, reports, and other media. The number of persons in the world who are engaged in generating and publishing scientific information has increased enormously. We are told that 90% or more of all the scientists that ever lived are living today. All this means an enormous amount of communication, in journals, bulletins, etc.

The mission of the scientist is to learn about and understand all natural phenomena. In the early beginnings a scientist worked alone, and, with his own hands, did everything that needed to be done, designing his own equipment, making his tools, constructing his apparatus, recording his observations, etc. In those days a scientist was concerned with all natural phenomena and knew all the science that existed. Subdivisions of science into the disciplines of astronomy, chemistry, biology, physics, geology, etc., came later. Our lone scientist maintained his own records. Occasionally he was able to communicate orally, or in writing, with a fellow scientist in his own or another country. As the number of scientists slowly increased, he became aware of the need to know what his fellow scientists were doing, and the volume of correspondence on scientific matters increased.

Then came scientific societies, for oral communication and for publication of papers in their proceedings, and journals for periodic publication. As time went on, individual scientists found that they were obliged to specialize more and more and to reduce their range of interest to smaller areas of science to be able to keep abreast of the knowledge being created. At this stage the individual scientist was still able to be in direct personal contact with substantially all of the original published papers in his restricted field, and he knew well all the relatively few journals he had to read.

However, with the further passage of time and the continued expansion in scientific output, the scientist found it increasingly difficult to know what papers in his field were appearing in what journals. Then came our abstract journals, which examined all primary scientific journals in the world in given fields and reported abstracts of the papers. In this way the individual scientist was directed to the original scientific papers in his field.

But the working time available to a scientist remains unchanged, and, in fact, may have decreased because of the proliferation of administrative procedures and accounting methods that we require in our modern ways of operation. The scientist has reached the point where the hours required to read and properly digest all of the original literature in his field will soon take up all his working time. The scientist can, of course, further subdivide his field of interest, and hold his own that way. But there is a limit to this continued splintering of science – learning more and more about less and less.

Such a situation is incompatible with our modern need for interdisciplinary investigations, since narrow specialists in two different disciplines may find it very difficult simply to communicate, let alone to work together in a fruitful manner to solve new problems requiring the experience and knowledge of two or more disciplines. Something different must be done to make it possible for the scientist today to assimilate fully the new knowledge being generated in his field.

B. CRITICAL TABLES OF STANDARD REFERENCE DATA

In principle, the solution is simple: We can interpose, between the original literature and the end-using scientist, a system of review and

*The material in this chapter was adapted from the author's previous publications in this field.[26-32]

appraisal by qualified experts of all the scientific information in given fields. Such reviews and appraisals would be performed for each given field, with the results of their work being made available through appropriate media.

This idea of having original information in the scientific literature reviewed and appraised for the end-using scientist is not new. In fact, in some measure, it has been going on for decades in the form of review articles covering specialized areas in given sciences. But these efforts at review and appraisal of the original scientific literature have in large part in the past been carried on sporadically.

The analysis and appraisal of scientific information is an intellectual task. Though it can be aided by machines, automatic devices, and high-speed computers, such work can be performed effectively and efficiently only by skilled scientists. In the long run, the scientific world can cope with the explosion of scientific information only if more scientists and technologists will dedicate themselves to the work of reviewing, correlating, and appraising scientific information.

Accurate experimental measurement is a costly enterprise, and unnecessary repetition of it is to be avoided. Therefore, science and industry need to have such numerical data collected, appraised, and formed into critical tables of standard reference data. The advance of science, technology, and industry rests heavily upon the quantitative scientific information we identify as critical tables of standard reference data.

For quantitative information of this kind, the entire communication process may be thought of as occurring in four steps: (a) the generation and publication of the data by the original investigator; (b) the appraisal of the data and the production of selected best values by competent experts to produce critical tables of standard reference data; (c) the dissemination of these critical tables in appropriate form; and (d) the use of the tables by the scientific and technical community.

If we can have adequate support for these critical tables of standard reference data produced by staffs of competent experts, we will have made a great step forward in the efficient utilization of our scientific manpower. Scientists in research laboratories will then have a maximum of time for their respective missions, fortified with the knowledge that they have at their disposal essentially all of the existing quantitative information in the literature in the form of standard reference data, critically prepared by experts.

The number of man-hours that could be saved in our laboratories would be incalculable. But what is more important, the numerical values available would be of much higher quality than can be produced by the sporadic effort of scientists primarily interested in other problems. This latter point is very important in the highly competitive technological world of today, where the precise control of temperature and pressure makes possible the conduct of industrial reactions and processes heretofore considered impossible.

C. SOME EARLY COMPILATIONS OF NUMERICAL DATA

Let us now refer to some early compilations of data.

The *Landolt-Bornstein Tables* began publication in Germany almost a century ago. The first edition of these Tables appeared in 1883 as one volume of 281 pages. The second, third, and fourth editions were also each in one volume, coming out in 1894 with 575 pages, in 1905 with 877 pages, and 1912 with 1,330 pages. The fifth edition, in eight volumes of about 7,500 pages, came out in the years 1923 to 1936. The sixth edition has had a total of 28 books with about 20,000 pages coming out in the years 1950 to 1970. Following the sixth edition, it was decided to alter the format of the *Landolt-Bornstein Tables* from complete general coverage to separate individual volumes covering specialized areas. Such volumes have appeared for nuclear physics and technology, magnetic properties, astronomy and astrophysics, atomic and molecular physics, crystal and solid-state physics, geophysics, etc.

The *Annual Tables of Constants and Numerical Data* had ten volumes appear in France in the years 1910 to 1930. Between 1936 and 1945, 40 installments appeared, covering the literature to 1939. In 1947, under the new name of *Tables of Selected Constants*, monographs on specialized areas have appeared. These include volumes on nuclear physics, spectroscopy, oxidation-reduction potentials, optical-rotatory power, semiconductors, and terpenoids.

The *Kaye and Laby Tables of Physical and Chemical Constants*, headquartered in the United Kingdom at the National Physical Laboratory, near London, have appeared in one volume, running through 13 editions, from 1911 to 1966.

The most notable example of an effort to provide critical tables of standard reference data is the *International Critical Tables of Numerical Data of Physics, Chemistry, and Technology.* The *International Critical Tables* resulted from discussions at the 1919 Conference of the International Union of Pure and Applied Chemistry at London. In 1923 the International Research Council, which was the predecessor of our present International Council of Scientific Unions (ICSU), gave its blessing to the project. For this project, the United States was assigned the financial and editorial responsibility. The National Academy of Sciences-National Research Council accepted the executive, editorial, and financial responsibilities for the United States, and, with the cooperation of industry, through the American Chemical Society and the American Physical Society, created a Board of Trustees to take charge of the financial and business administration and a Board of Editors to supervise the preparation of the tables proper. The entire enterprise was made possible by a fund of $200,000, contributed largely by approximately 200 industrial conpanies of the U.S. This famous collection of numerical data was the result of cooperative efforts by 408 scientists in 18 different countries. Eight volumes with a total of 3,819 pages were published in the years 1926 to 1933. This was the broadest single compilation job in history. It was hoped that the *International Critical Tables* would become a continuing operation, with revisions from time to time, but, unfortunately, the Editor in Chief, Edward W. Washburn, died in 1934, and this continuity never came to pass. The importance of the *International Critical Tables* to the science and technology of the world is evident from the fact that a significant number of sets of these tables are being sold today, a half century after their original issue.

D. COMPILATIONS OF CHEMICAL THERMODYNAMIC DATA

There is one class of data where critical appraisal and selection of best values is particularly important. This is the class of thermodynamic data. Here, one is required to have values which are not only consistent among themselves but also consistent with the relations of thermodynamics, physics, and chemistry. Because of the particular importance attached to these tables, one may briefly summarize what has been done in this area, apart from what has appeared in the *International Critical Tables,* the *Landolt-Bornstein Tables,* and the *Annual Tables* referred to above.

One of the earliest tables of heats of formation of chemical compounds was put together by Julius Thomsen of Denmark, published from 1882 to 1886.[1]

In 1923 Lewis and Randall of the U.S. published their tables of the heats and free energies of some chemical substances.[2]

In 1936 appeared the book of Bichowsky and Rossini[3] on the *Thermochemistry of the Chemical Substances.* This work revised and extended considerably the compilation by Bichowsky which appeared in Volume V of the *International Critical Tables.*[4]

Meanwhile, over the years 1934 to 1962 appeared the compilations of K. K. Kelley of the U.S. Bureau of Mines, on the thermodynamic properties of metallurgically important substances.[5]

In 1944 appeared the first tables of thermodynamic properties of hydrocarbons and related compounds of the American Petroleum Institute Research Project 44, under Rossini and, later, Zwolinski.*[6] These loose-leaf tables came out in book form in 1947[7] with a second volume in 1953.[8] Essentially all thermodynamic properties of hydrocarbons and related compounds are covered in these tables, which constitute the most comprehensive work in this field ever published.

In 1952 appeared the large National Bureau of Standards Circular 500, on *Selected Values of Chemical Thermodynamic Properties,* by Rossini, Wagman, Evans, Levine, and Jaffe.[9] This was a revision and extension of the book of Bichowsky and Rossini.

In 1955 the first of the loose-leaf data sheets on *Selected Values of the Thermodynamic Properties*

*Operated at the National Bureau of Standards, 1942–1950, at the Carnegie Institute of Technology (now Carnegie-Mellon University), 1950–1961, and at Texas A & M University, 1961 to the present. Frederick D. Rossini was director of this project from 1942 to 1960 and Bruno J. Zwolinski from 1960 to the present. Information on these loose-leaf data sheets of physical, thermodynamic, and spectral properties of hydrocarbons and related compounds may be obtained from B. J. Zwolinski, Thermodynamics Research Center, Texas A & M University, College Station, Texas 77843.

of Metals and Alloys appeared, under Hultgren at the University of California at Berkeley.[10]

In 1956 appeared the first of the loose-leaf data sheets of the Manufacturing Chemists Association Research Project under Rossini and Zwolinski, later becoming the Thermodynamics Research Center Data Project, on *Selected Values of Properties of Chemical Compounds.* *[11]

In 1960, under Stull at the Dow Chemical Company, came the first of the loose-leaf tables of the thermodynamic properties of properties of propellants and related substances, under the JANAF label.**[12]

In 1962, in the U.S.S.R., appeared the volume entitled *Thermodynamic Properties of Chemical Substances,* by Glushko, Gurvich, Medvedev, and others, on propellants and related substances.†[13]

In 1965, in the U.S.S.R., the volume entitled *Thermodynamic Constants of Substances,* by Glushko, Medvedev, Gurvich, and others, appeared.†[14]

Something of the present status of the continuing work is summarized as follows [31]

The American Petroleum Institute Research Project 44 is continuing its work in the Thermodynamics Research Center at Texas A & M University,[6] under Zwolinski and co-workers, dealing with hydrocarbons and related compounds. Also associated with this work is the Thermodynamics Research Center Data Project,[11] under Zwolinski and co-workers, dealing with chemical substances other than hydrocarbons, funded largely by the National Bureau of Standards Office of Standard Reference Data. The data of these two projects include more than 16,000 loose-leaf data sheets, of which nearly 3,000 relate to thermodynamic properties.

At the National Bureau of Standards, work is nearing completion on the revision of NBS Circular 500 on *Selected Values of Chemical Thermodynamic Properties,* under Wagman, Evans, and co-workers. NBS Technical Notes 1, 2, 3, 4, 5, 6, and 7 have appeared, with Technical Note 8 due before 1975. These eight booklets will cover, in the standard order of arrangement,[3,9] all the chemical elements and their compounds except the compounds containing the alkali metals, and will have over 600 pages of numerical data. The section containing compounds of the alkali metals will be the largest – nearly 200 pages. The final complete book will include all references.

The project on *Selected Values of the Thermodynamic Properties of Metals and Alloys,* by Hultgren and co-workers at the University of California at Berkeley, continued to the fall of 1972. The final volume of this work covers 66 elements on 435 pages and 222 alloy systems on 779 pages, making a total of 288 systems on 1,214 pages.[10] The bibliography comprises about 8,800 entries. The file of references, the search system, and the microfiche collection developed in this work are maintained for the U.S. Atomic Energy Commission, sponsor of this work, in the custody of Leo Brewer, Inorganic Materials Research Division, Lawrence Laboratory, University of California at Berkeley.

The JANAF Thermochemical Tables Project at the Dow Chemical Company, Midland, Michigan, now under M. W. Chase, is continuing its work under the sponsorship of the U.S. Air Force Office of Scientific Research. As of the beginning of 1974, 40 supplements have been issued covering 35 elements and 1,245 species.[12] A new bound volume will be issued in 1974, giving new and revised data sheets, and will serve as a supplement to the bound volume entitled *JANAF Thermochemical Tables,* which covered 1,099 species.[15] In this work emphasis is on the smaller species, including ions, likely to exist at high temperatures and important in propellant and combustion processes.

*Manufacturing Chemists Association Research Project; operated at Carnegie Institute of Technology (now Carnegie-Mellon University), 1955 to 1961, and at Texas A & M University, 1961 to the present. In 1963 the Manufacturing Chemists Association support was replaced by support from the National Bureau of Standards Office of Standard Reference Data, and the name of the project was changed to Thermodynamics Research Center Project. Frederick D. Rossini was director of this project, 1955 to 1960, and Bruno J. Zwolinski, 1960 to the present. Information on these loose-leaf data sheets may be obtained from B. J. Zwolinski, Thermodynamics Research Center, Texas A & M University, College Station, Texas 77843.

**Operated at the Dow Chemical Company, Midland, Michigan, from 1959 to the present. Daniel R. Stull was director of this project, 1959 to 1969, Harold Prophet, 1969 to 1972, and Malcolm W. Chase, 1972 to the present. Information on these loose-leaf data sheets may be obtained from M. W. Chase, Thermal Laboratory, Dow Chemical Company, Midland, Michigan 48640.

†Information on these tables may be obtained from the U.S.S.R. Academy of Sciences, and the All-Union Institute of Scientific and Technological Information, Moscow, U.S.S.R.

The project in the U.S.S.R. on the *Thermodynamic Properties of Chemical Substances* is continuing under Glushko, Gurvich, Medvedev, and co-workers, supported by the Academy of Sciences of the U.S.S.R. The compounds selected are those important in propellant and combustion processes. The third volume contains data on the thermodynamic properties of about 800 substances from 100K to very high temperatures. Plans for work through 1975 were scheduled for this project.[13]

The project in the U.S.S.R. on the *Thermodynamic Constants of Substances* is continuing under Glushko, Medvedev, Gurvich, and co-workers, supported by the Academy of Sciences of the U.S.S.R. Four volumes have so far been published in this work, which uses the standard order of arrangement and follows the pattern of NBS Circular 500, including all inorganic compounds and organic compounds with 1 and 2 carbon atoms. The fourth volume contains data on 4,500 substances, in 920 pages, with 7,000 references. The fifth volume covers about 2,500 substances in about 500 pages, with 3,500 references. The sixth volume covers about 2,400 substances, with 8,000 references.[14]

The compilations on thermodynamic properties referred to above are those which are broad based and constitute enterprises of considerable magnitude. In addition, there are many other compilation jobs on thermodynamic properties being performed in many smaller and specialized areas of science. Reference to these is made in the *CODATA International Compendium of Numerical Data Projects,*[16] which is discussed later in this chapter, as well as the National Research Council's (U.S.) *Continuing Numerical Data Projects – A Survey and Analysis.*[17]

E. OTHER COMPILATIONS OF NUMERICAL DATA

Compilations of numerical data in many other areas of science and technology are being produced in the United States and in other countries of the world. These compilations cover many different specialized areas of chemistry, physics, thermophysics, nuclear science, space science, astronomy, etc.

The National Research Council's (U.S.) *Continuing Numerical Data Projects – A Survey and Analysis*[17] and the *CODATA International Compendium of Numerical Data Projects*[16] give detailed information of such projects in the following areas of science:

1. Nuclear properties: nuclear data; nuclear tables; nuclear constants; tables of isotopes; nuclear radii; reactor physics constants; neutron cross sections; charged-particle cross sections; decay schemes for radioactive nuclei; energy levels of nuclides; photonuclear data.
2. Atomic and molecular properties: atomic energy levels; atomic transition probabilities; X-ray wavelengths; X-ray energy levels; diatomic molecular spectra and energy levels; molecular structure; molecular vibration frequencies; atomic and molecular ionization processes; dipole moments.
3. Spectral properties: infrared, ultraviolet, Raman, mass, microwave, nuclear magnetic resonance, electronic, and other atomic and molecular spectra.

Also covered are projects concerned with the following solid-state properties: crystallographic, mineralogical, electrical, magnetic, superconductors.[16,17]

In addition to the foregoing, projects involving the following properties are reviewed in these two volumes:[16,17] thermophysical and transport properties, properties of solutions, chemical reaction rates, gas chromatographic data, and optical rotatory properties.

F. HANDBOOKS OF NUMERICAL DATA

Reference is made here to handbooks of numerical data into which the data originally compiled by the experts mentioned in the preceding section eventually find their way. Some of these handbooks are prepared for a special clientele, while others are for a broad, general audience consisting of those who need reliable but not necessarily the most accurate data currently available in a given area.

The *CODATA International Compendium of Numerical Data Projects*[16] lists the following "Desk Handbooks for Broad Fields of Science":

Physical and Chemical Sciences

Handbook of Chemistry and Physics, Weast, R. C., Ed., CRC Press, Cleveland, Ohio.
Tables of Physical and Chemical Constants and Some Mathematical Functions, Kaye, G. W. C. and Laby, T. H., Longmans, Green, London.
Taschenbuch für Chemiker und Physiker, D'ans Lax, Lax, E. and Synowietz, C., Springer, Berlin.
Handbuch des Chemikers, Nikolski, B. P., Vol. I, Allgemeines; Vol. 2, Die Chemischen Elements; Vol. 3, Chemisches (Gieichgewicht und Kinetik, VEB Verlag Technik, Berlin.
Handbook of Chemistry, Lange, N. A., Ed., McGraw-Hill, New York.
Handbook of Chemistry (Kagaku Binran), Chemical Society of Japan, Maruzen Publishing Co., Tokyo.
Concise Handbook on Physicochemical Values, Mischenko, K. P., Ed., Khemia Publishing House, Moscow.
Physical Property Values of Substances (Bussei Teisu), Japan Society of Chemical Engineers, Chairman, Editorial Committee, Kazuo Sato, Maruzen Publishing Co., Tokyo (annual monographs for chemical engineers, 60 journals abstracted, noncritical).
Handbook of Inorganic Chemistry (Muki Kagaku Binran), Gihodo Publishing Co., Tokyo.
Handbook of Organic Chemistry (Yuki Kagaku Binran), Gihodo Publishing Co., Tokyo.
American Institute of Physics Handbook, Gray, D. E., Ed., McGraw-Hill, New York.
Physikalisches Taschenbuch, Ebert, H., Vieweg & Sohn, Braunschweig, Germany.
Smithsonian Physical Tables, Forsythe, W. E., Ed., Smithsonian Institution, Washington, D.C.

Biological Sciences

Handbook of Biochemistry, Selected Data for Molecular Biology, Sober, H. A., Ed., Chemical Rubber Co., Cleveland.
Biology Data Book, Atman, P. L. and Dittmer, D. S., Federation of American Societies of Experimental Biology, Washington, D.C.
Biochemists' Handbook, Long, C., Ed., D. Van Nostrand Co., Princeton, N.J.

Earth Sciences

Handbook of Physical Constants, Clark, S. P., Jr., Ed., The Geological Society of America, Inc., 231 East 46th Street, New York, N.Y. 10017.
Physical Properties of Rocks under Normal and Standard Temperature and Pressure, by a group of authors of the Institute of the Physicochemical Basis for Treating Raw Materials, Nauka Publishing House, Moscow.
Elastic Properties of Rock-forming Minerals and Rocks, Bellikov, P. B. et al., Nauka Publishing House, Moscow.
Reference Book on Physicochemical Values for Geochemists, Naumov, G. B. et al., Nauka Publishing House, Moscow.
Smithsonian Meteorological Tables, List, R. J., Ed., Smithsonian Institution, Washington, D.C.
International Meteorological Tables, Letestu, S., Ed., World Meteorological Organization, Geneva.

Nuclear Sciences

Group Constants for Nuclear Reactor Calculations, Abagyan, L. B., Bazazyants, N. L., Bondarenko, I. I., and Nikolaev, M. N., revised American edition, Power Physics Institute, Institute of Atomic Energy, Academy of Sciences of the U.S.S.R., Plenum Press, 1964.
Compendium of Thermal-Neutron Capture γ Ray Measurements, Part I, $Z \leqslant 46$, Bartholomew, G. A. (Atomic Energy of Canada, Ltd., Chalk River, Ont.), Daveika, A., Eastwood, K. A., Monaro, S., Groshev, L. V., Demidov, A. M., Pelekhov, V. I., and Sokolovskii, L. L., Nucl. Data, Sect. A, 3, 367, 1967.
Tables des Isotopes, Pannetier, R., Vol. 2, Maisonneuve, S. A., Moulins, France, 1965.
Concise Handbook for the Engineering Physicist: Nuclear Physics, Atomic Physics, Federov, N. D., BR-SOV/5425, Atomizdat Publishing House, Moscow, 1961.
Tables of Neutron Resonance Parameters and Neutron Resonance Materials (for structural materials, Golashvily, T. B. and Elagin, Yu, P., Publishing House for State Standards, Moscow.
Systematic Presentation of Isotopes and Reference Diagrams for Nuclides, Selinov, I. P., Publishing House for State Standards, Moscow, 1968.

The *CODATA International Compendium of Numerical Data Projects* also gives information on handbooks for specialized areas of science, including nuclear properties, spectroscopic properties, solid-state properties (crystallographic, mineralogical, electrical, and magnetic), and thermodynamic and transport properties.[16]

G. DEVELOPMENTS IN THE UNITED STATES ON DATA FOR SCIENCE AND TECHNOLOGY

Most of the compilation work prior to World War II was performed mainly by scientists in their spare time. But with the enormous increase in numerical data generated it became impossible for such work to continue to be carried on by individuals as a private enterprise on their own time.

By 1955, formal compilation projects were being carried on in several countries in response to the urgent need of science and technology for more up-to-date numerical data in support of the research programs being accelerated everywhere.

In the United States, which had been the headquarters of the *International Critical Tables,* special situations took place. The National Research Council established a Committee on Tables of Constants in the early Forties. Although this Committee did consider it desirable to have a revision of the *International Critical Tables*, it saw no ready solution to that problem.

In 1955, the National Research Council Committee on Tables of Constants concluded that there was no hope of repeating the work of the *International Critical Tables*, as one large compilation job, for the following reasons:

1. Since the appearance of the *International Critical Tables* in 1926 to 1930, the fields of chemistry, physics, engineering, technology, and other sciences had expanded in size and created new areas, requiring many times more and newer data than before.

2. The precision of measurement in science, and the precision of manufacturing in industry, have been pushed to another magnitude, requiring more accurate data of greater precision.

3. A rough estimate indicated that an adequate and complete revision and extension of the *International Critical Tables* would be a job several hundred times as great as the original job and would require nearly $100,000,000.

4. From 1938 to 1955, a number of large data-compiling projects operating on a continuing basis had come into existence in the United States, involving total annual expenditures of about $1,000,000. These projects included the American Petroleum Institute Research Project 44 on hydrocarbons and related compounds, the Manufacturing Chemists Association Research Project on chemical compounds, the Nuclear Data Project of the United States Atomic Energy Commission, the compilation work on metallurgically important compounds of the U.S. Bureau of Mines, the Thermophysical Properties Center at Purdue University, and several compilation projects on thermochemical, thermodynamic, and other properties at the National Bureau of Standards.

Any new undertaking of this kind must provide for one essential element lacking in the *International Critical Tables,* namely firm provision for continuity. The National Academy of Sciences-National Research Council decided that a new plan for providing science and industry in the United States with continuing up-to-date critical tables of data was needed. There was then established in the National Research Council an Office of Critical Tables, with the following responsibilities:

a. To survey the needs of science and industry for critical tables of numerical data for chemistry, physics, earth sciences, and engineering;

b. To stimulate and encourage, and expand as appropriate, existing critical data-compiling projects;

c. To promote uniform editorial policy and procedures, and high standards of quality;

d. To provide a directory of continuing critical data-compiling projects;

e. To assist in the establishment of needed critical data-compiling projects for new scientific areas, in responsible institutions, manned by qualified investigators, with financial support provided by appropriate organizations having a specific or general interest in the data.

The Office of Critical Tables of the National Research Council began operating late in 1957 with Guy Waddington as director. In the first 6 years, significant progress was made in carrying on the objectives listed under Items a, b, c, and d above. But, for Item e the establishment of new data-compiling projects for new scientific areas required considerable funding and little progress was made on this objective.

However, in 1963, a new impetus was given to this last objective when funding was made available by the U.S. Government to the National Bureau of Standards to take over the operation for the Government of a National Standard Reference Data Program.

The National Bureau of Standards established an Office of Standard Reference Data to carry on the work required for the new program and to create a National Standard Reference Data System. Edward W. Brady was the chief of this Office and, subsequently, on assuming higher administrative duties at the National Bureau of Standards in 1968, was succeeded by the present chief of that Office, David R. Lide, Jr. For this work the funding by the government has been of the order of $2.5 million per year. Support has been provided at the National Bureau of Standards for a number of in-house data-compiling projects and other work related to data-compiling and data-retrieving operations, and for a large number of projects in other organizations in the U.S. Publications arising from work supported in this program are published as books or bulletins in the series labeled "NSRDS-NBS" and also in the newly established, 1972, *Journal of Physical and Chemical Reference Data,* of which Lide is the editor.[18,19]

The NRC Office of Critical Tables had been helpful in the establishment of the NBS Office of Standard Reference Data, worked closely in its formation, and provided an advisory committee to monitor its on-going program. In 1969, when Waddington retired, the NRC Office of Critical Tables was reduced in operations to become the NRC Numerical Data Advisory Board,[20] principally to function as advisor to the NBS Office of Standard Reference Data and as the liaison office for the U.S. with CODATA (discussed in the following section).

H. INTERNATIONAL DEVELOPMENTS ON DATA FOR SCIENCE AND TECHNOLOGY

With this background, let us now review briefly how CODATA, the Committee on Data for Science and Technology, under ICSU, the International Council of Scientific Unions, was established. Early in 1964, when it was abundantly clear to all that the problem of data for science and technology was truly an international one, it was suggested by Rossini and Waddington, through the National Research Council (U.S.) Office of Critical Tables, of which they were chairman and director, respectively, that ICSU might take the lead in providing international coordination and guidance in this field. In June, 1964, ICSU established a Working Group, consisting of Harrison Brown, Chairman, F. D. Rossini (U.S.), Sir Gordon Sutherland (U.K.), Wilhelm Klemm (Germany), V. A. Kirillin (U.S.S.R.), and B. Vodar (France) as members and G. Waddington (U.S.) as a resource person. The Working Group met late in 1964 and formulated a favorable recommendation to ICSU. In April, 1965, the recommendation was approved. At a meeting in September, 1965, the Working Group prepared a constitution and a list of members – unions and countries. In January, 1966, the General Assembly of ICSU, with Sir Harold Thompson (U.K.) in the chair, approved the establishment of a Committee on Data for Science and Technology, with a constitution and initial membership. (Waddington later coined the acronym CODATA for the Committee.)

For its first 2 years, CODATA had its Central Office at the National Research Council in Washington, D.C., with Waddington as director. As of July 1, 1968, the Central Office of CODATA was moved to Frankfurt, Germany, with Christoph Schafer as director. The author served as president of CODATA, 1966 to 1970, with representatives of the principal other countries on the Executive Committee: Sutherland (U.K.); Klemm (Germany); Styrikovich (U.S.S.R.); Kotani (Japan); Vodar (France).

In 1970 the Central Office of CODATA had the following staff: executive director; two scientists, one of whom served as editor; and two secretaries. Vodar served as president of CODATA from 1970 to 1974. In 1973 the structure of CODATA was reorganized, with the Central Office being reduced in staff and headed by an executive secretary. The post of secretary-general of CODATA was created, with increased responsibilities, with Edgar F. Westrum, Jr. (U.S.), taking this assignment.

CODATA includes in its membership about two thirds of the 16 scientific unions comprising ICSU, and about 15 national members. The latter are countries that have one or more significant programs on compilation under way, and that pay annual dues to CODATA through their national adhering body to ICSU. In 1970–72, the budget of CODATA was of the order of $100,000 per year.

All of the countries that are members of CODATA have each established a national committee for CODATA, and in several cases, including the U.S., U.K., U.S.S.R., Germany, France, and Japan, have established national programs dealing with data for science and technology.

At its start, CODATA was given the following assignments on a worldwide basis:

a. To ascertain, through the unions and appropriate national bodies, what data-compiling work is going on and what the needs are;

b. To achieve coordination among, and provide guidance for, data-compiling projects;

c. To encourage support for data-compiling projects by appropriate private, governmental, and intergovernmental agencies;

d. To encourage the use of internationally approved constants, units, and symbols, and, when desirable, uniform editorial policy and procedures;

e. To produce a directory-survey of continuing data-compiling projects and related work;

f. To encourage and coordinate research on new forms for preparing and distributing critically evaluated numerical data.

To carry on its work, CODATA decided on several missions, as follows:

a. To prepare, publish, and maintain up to date an International Compendium of Numerical Data Projects;

b. To issue a CODATA Newsletter, semiannually, for distribution to all interested scientists, to give information about CODATA activities, data compilations, etc.;

c. To hold, biennially, International Conferences on the Generation, Collection, Evaluation, and Dissemination of Numerical Data for Science and Technology;

d. To establish CODATA Task Groups, to work on special problems necessary for compilation work on data for science and technology;

e. To issue a CODATA Bulletin, containing reports of the several CODATA Task Groups.

CODATA has proceeded expeditiously on these missions:

a. The *CODATA International Compendium on Numerical Data Projects* appeared in publication late in 1969. The material for the volume was prepared by Waddington and his staff in Washington, beginning in 1967 and concluding in early 1969. The copy was then sent to CODATA headquarters in Frankfurt, Germany, where Schafer arranged for its prompt publication by Springer.[16]

b. The *CODATA Newsletter* has appeared as planned.[21]

c. CODATA International Conferences on the Generation, Collection, Evaluation, and Dissemination of Numerical Data for Science and Technology have been held every 2 years: in 1969 in Germany, near Frankfurt; in 1970 in the United Kingdom, at St. Andrews; in 1972 in France, at Le Creusot; in 1974 in the U.S.S.R., near Erivan, Armenia.[22]

d. The CODATA Task Groups established include the following:

1. Task Group on Computer Use – concerned with coordination and compatibility of equipment and programs;

2. Task Group on Fundamental Constants – concerned with developing and maintaining a recommended list of fundamental constants for worldwide use;

3. Task Group on Key Values for Thermodynamics – concerned with developing agreement on values for "key" substances involved in thermodynamic tables and calculations;

4. Task Group on Data for Chemical Kinetics – concerned with developing procedures and standards for presenting and compiling data for chemical kinetics;

5. Task Group on Publication of Data in the Primary Literature;

6. Task Group on Accessibility and Dissemination of Data.

e. The *CODATA Bulletin*[23] has been issued at intervals, carrying reports of CODATA Task Groups.

Reports of the CODATA Task Group on Computer Use have appeared in the *CODATA Bulletin*.[24] A symposium sponsored by this

CODATA Task Group was held in July, 1973, at Freiburg i Breisgau, Germany, with the title "Man-Machine Communication for Scientific Data Handling."

The report of the CODATA Task Group on Fundamental Constants, giving the 1973 recommended values, has been discussed in detail in Chapter 12.

Reports of the CODATA Task Group on Key Values for Thermodynamics have been published in the CODATA Bulletin Numbers 2, 6, 7, and 10.[25] A new report, Part IV, is appearing in 1974.

The first report of the Task Group on Publication of Data in the Primary Literature appeared in 1973 and is discussed in the following chapter.

As of this writing, reports of the other CODATA Task Groups are in preparation.

I. DISCUSSION

As may be noted from the material presented in this chapter, two important changes in course, unrelated chronologically, occurred in the area of data for science and technology: (a) recognition of, and action on, the problem of data for science and technology as an international enterprise came with the birth and demise of the International Critical Tables organization, 1920–1934, and later with the establishment in 1966 by the International Council of Scientific Unions, ICSU, of the Committee on Data for Science and Technology, CODATA; (b) about 1940, it was recognized that, except for some specialized areas of limited size, the work of the collection, analysis, calculation, and compilation of critical tables of standard reference data was no longer one that could be done by individual scientists in their spare time in evenings and weekends, but rather by full-time scientific specialists dedicating their talents to such work.

It is hoped that CODATA will evolve into a viable enterprise that will truly coordinate and encourage the production of needed working tabulations of critical tables of standard reference data for the scientific and technical community of the world.

The author has published a number of reports dealing with the problem of data for science and technology.[26–32]

Finally, the author makes some general comments about critical tables of standard reference data as he would like to see them throughout the world:

1. The tables should cover all substances and properties for which any information is available.
2. The tables should have a chemically oriented standard order of arrangement of the compounds, with the arrangement, language, and symbols readily understood by workers in all countries, in any language.
3. The tables should be fully self-consistent, including consistency with all the physical relations involved and with the internationally recommended values of the fundamental constants and atomic masses.
4. The tables should be produced by competent scientists of high capabilities, with adequate rewards for them.
5. The tables should be produced on a continuing basis and maintained up to date by revision at appropriate intervals.
6. The work should be adequately supported, with resources provided by both government and industry as partners in this important work.
7. The tables should be easily available at reasonable cost in any part of the world.

In making assignments of areas of responsibility, we should take advantage of, and utilize to the fullest extent, the capabilities of all workers interested in, and willing to work on, the preparation of critical tables, wherever they happen to be in the world.

Also, in making assignments of areas of responsibility to different projects, we need to avoid significant duplication of effort. At the same time, however, we need enough overlapping to ensure that we get the benefits of competitive comparison of the quality of the work performed by different projects.

A point important in the entire business of producing critical tables of standard reference data is that the data must be on the shelf at the time of need. We cannot wait for the preparation until there is an important need for a specific lot of data. It will then be too late. Many important industrial enterprises have been subject to very costly delays because of the lack of a few data of the right kind at the right time. Similarly, many research investigations have suffered because of the lack of a few needed data. In a related vein, unreliable data have been known to lead to great losses in industrial enterprises and to much wasted effort in research investigations.

For a successful and fruitful operation in the production of critical tables of standard reference data, we must arrange for adequate coordination of all projects on national and international levels, and we must maintain a system of communication in good working order to provide channels of contact and transfer of appropriate information at all levels for all workers in all sciences in all countries.

The author is indebted for discussions on this problem to Guy Waddington, retired director of the Office of Critical Tables, National Research Council (U.S.), and first executive director of CODATA.

J. REFERENCES

1. **Thomsen, J.,** *Thermochemische Untersuchungen,* Vol. 1–4, Barth, Leipzing, 1882, 1882, 1883, 1886.
2. **Lewis, G. N. and Randall, M.,** *Thermodynamics and the Free Energy of Chemical Substances,* McGraw-Hill, New York, 1923.
3. **Bichowsky, F. R. and Rossini, F. D.,** *Thermochemistry of the Chemical Substances,* Reinhold, New York, 1936.
4. **Bichowsky, F. R.,** in *International Critical Tables,* Vol. 5, Washburn, E. W., Ed., McGraw-Hill, New York, 1929.
5. **Kelley, K. K.,** *Contributions to the Data on Theoretical Metallurgy,* Bulletins of the U.S. Bureau of Mines, No. 383, 384, 393, 406, 542, 584, 592, 601, U.S. Government Printing Office, Washington, D.C., 1935 to 1962.
6. American Petroleum Institute Research Project 44, Selected Values of Properties of Hydrocarbons and Related Compounds.
7. **Rossini, F. D., Pitzer, K. S., Taylor, W. J., Ebert, J. P., Kilpatrick, J. E., Beckett, C. W., Williams, M. G., and Werner, H. G.,** *Selected Values of Properties of Hydrocarbons,* National Bureau of Standards Circular 461, U.S. Government Printing Office, Washington, D.C., 1947.
8. **Rossini, F. D., Pitzer, K. S., Arnett, R. L., Braun, R. M., and Pimentel, G. C.,** *Selected Values of Physical and Thermodynamic Properties of Hydrocarbons and Related Compounds,* Carnegie Press, Carnegie Institute of Technology, Pittsburgh, 1953.
9. **Rossini, F. D., Wagman, D. D., Evans, W. H., Levine, S., and Jaffe, I.,** *Selected Values of Chemical Thermodynamic Properties,* National Bureau of Standards Circular 500, U.S. Government Printing Office, Washington, D.C., 1952.
10. **Hultgren, R. R. et al.,** *Selected Values of Thermodynamic Properties of Metals and Alloys,* University of California Press, Berkeley, Cal.
11. Selected Values of Properties of Chemical Compounds, Manufacturing Chemists Association Research Project.
12. JANAF Thermochemical Tables Project, under the sponsorship of the U.S. Air Force Office of Scientific Research, as a Joint Army-Navy-Air Force Project.
13. **Glushko, V. P., Gurvich, L. V., Medvedev, V. A., et al.,** Thermodynamic Properties of Chemical Substances, project under the U.S.S.R. Academy of Sciences, Moscow.
14. **Glushko, V. P., Medvedev, V. A., Gurvich, L. V., et al.,** Thermodynamic Constants of Substances, project under the U.S.S.R. Academy of Sciences, Moscow.
15. **Stull, D. R. and Prophet, H.,** *JANAF Thermochemical Tables,* 2nd ed., NSRDS-NBS-37, U.S. Government Printing Office, Washington, D.C., 1971.
16. *CODATA International Compendium of Numerical Data Projects,* Springer, Heidelberg, New York, 1969. Additional information may be obtained from the CODATA Central Office, ICSU, 51 Blvd. de Montmerency, F-75016 Paris, France.
17. *Continuing Numerical Data Projects – A Survey and Analysis,* 2nd ed., Publ. 1463, National Academy of Sciences, 2101 Constitution Avenue, Washington, D.C., 1966.
18. National Bureau of Standards Office of Standard Reference Data, David R. Lide, Jr., chief, National Bureau of Standards, Washington, D.C. 20234.
19. *Journal of Physical and Chemical Reference Data,* Lide, D. R., Jr., Ed., National Bureau of Standards, Washington, D.C. 20234. Subscription Office: American Chemical Society, 1155 16th St. N.W., Washington, D.C. 20036.
20. Numerical Data Advisory Board, H. Van Olphen, executive secretary, National Research Council, Washington, D.C. 20418.
21. *CODATA Newsletter,* CODATA Central Office, ICSU, 51 Blvd. de Montmorency, F-75016 Paris, France.

22. CODATA International Conferences. Information on the programs may be obtained from the CODATA Central Office, ICSU, 51 Blvd. de Montmorency, F-75016 Paris, France.
23. *CODATA Bulletin,* CODATA Central Office, ICSU, 51 Blvd. de Montmorency, F-75016 Paris, France.
24. Reports of the CODATA Task Group on Computer Use (Gordon Black, chairman; R. N. Jones, secretary), *CODATA Bulletin,* CODATA Central Office, ICSU, 51 Blvd. de Montmorency, F-75016 Paris, France.
25. Reports of the CODATA Task Force on Key Values for Thermodvnamics (S. Sunner, chairman), *CODATA Bulletin,* CODATA Central Office, ICSU, 51 Blvd. de Montmorency, F-75016 Paris, France.
26. **Rossini, F. D.,** *Res. Management,* 10, 107, 1967.
27. **Rossini, F. D.,** *J. Chem. Doc.,* 7, 2, 1967.
28. **Rossini, F. D.,** Report of a Discussion Meeting for Science and Technology, Royal Society, London, 1967, 8.
29. **Rossini, F. D.,** *CODATA Newsl.,* 1, 2, 1968.
30. **Rossini, F. D.,** *J. Chem. Doc.,* 10, 261, 1970.
31. **Rossini, F. D.,** *Rev. Roum. Chim.,* 17, 267, 1972.
32. **Rossini, F. D.,** *J. Chem. Eng. Data,* 18, 113, 1973.

Chapter 14

PRESENTATION AND ANALYSIS OF NUMERICAL DATA IN THE SCIENTIFIC LITERATURE

A. INTRODUCTORY COMMENTS

In 1958 the author, a member of the executive committee for the then Office of Critical Tables of the National Research Council, prepared a report dealing with the presentation and analysis of numerical data in the scientific literature for distribution to the members of the committee and the clientele of the NRC Office of Critical Tables.[1] The material in this chapter comes largely from that report, supplemented with suggestions from the 1972 report of a subcommittee of the Commission on Thermodynamics and Thermochemistry of the International Union of Pure and Applied Chemistry[2] and from the 1973 report of the CODATA Task Group on Publication of Data in the Primary Literature.[3]

In this chapter we discuss the responsibility of the investigator in recording exactly what was measured, with what standards and references, and the responsibility of the reviewer and appraiser in retrieving from the literature the actual facts with a maximum of accuracy and precision.

Observation and measurement are the backbone of science. Numerical data arise from observation and measurement. Therefore, numerical data constitute the lifeblood of science. The precision and accuracy of measurement have increased enormously with advances in science. Over the past century, for example, the precision of measurement of length has increased by a factor of approaching one million. The experimenter needs to make the record of his work as useful as possible by utilizing all the devices at his disposal to attain high accuracy. Today, much of the numerical data appearing in the literature is of very high accuracy and precision. In the review and appraisal of such data it is important that all of the accuracy and precision be preserved in the transition from the original record to the final compilation.

B. RESPONSIBILITY OF THE INVESTIGATOR

Basically, the responsibility of a scientific investigator is to record for posterity in science, as well as for his contemporary scientists, exactly what he measured, what reference instruments or reference substances he used, how the experimental observations were reduced, what assumptions, if any, were involved in the reduction of the data, and what physical constants, atomic masses, etc., he coupled with his own observations to produce the final value or values he reported in the literature.

Unless the investigator is conscientious in recording all information that is relevant in the investigation, undesirable outcomes occur: (a) the appraiser will be unable to recover from the record the accuracy and precision that may have been present in the work; (b) the appraiser may be mistakenly led into crediting the investigator with an accuracy, or precision, or both, far greater than actually existed. In the first case, the work will likely be repeated unnecessarily at a later date, usually at a significant cost; in the second case, the user of the data will be led into serious error, usually at relatively great cost, whether the application of the data is in the laboratory or in the pilot plant.

Depending upon the property of the system (chemical substance) that is being subjected to investigations, the following are some of the measured quantities that may be involved:

1. Purity of the substance
2. Temperature
3. Pressure
4. Interval of time
5. Electric power
6. Length
7. Volume
8. Mass

Some comments on the foregoing follow.

With regard to the purity of the substance being studied, the investigator has a prime responsibility to establish that the amount and nature of the impurity in the substance are such as to be not significant within the limits of uncertainty of his measurements. The purity can be established by a reliable source that provides a certificate of purity and description of the method of determining the amount of impurity. The amount of impurity that can be tolerated will depend upon the property being measured and the nature of the impurity. For example, an impurity of 1 mol% of 2,4-dimethylhexane in the substance 2,5-dimethylhexane will affect the normal boiling point of the latter by about 0.003°C, the refractive index (n_D at 25°C) by about 0.00003, the density (g/cm^3) by about 0.00007, and the freezing point (in air at 1 atm) by 0.214°C.[4] Thus, the freezing-point behavior is seen to be a sensitive and accurate measure of the amount of impurity in a given substance when that impurity is liquid soluble and solid insoluble.[5-7] From the foregoing, one can easily calculate the amount of the impurity of 2,4-dimethylhexane in 2,5-dimethylhexane that can be tolerated for given uncertainties of measurement of the several properties.

In the measurement of the heat of combustion of 2,5-dimethylhexane, one can tolerate 10 mol% of impurity of 2,4-dimethylhaxane without affecting the result by more than 0.01%. But it is to be noted that such calculations are very much dependent on both the nature and the amount of impurity. All of this emphasizes the need to know the nature as well as the amount of impurity in a given substance in order to be able to calculate the tolerances for the required accuracy of the given measurement.

In measurements involving a given chemical reaction, as in measuring the heat energy associated with a given chemical reaction, it is necessary to establish not only the purity of each of the reacting substances, but also what we may call the "purity of the reaction," by appropriate analyses of the products of the reaction.

Those kinds of impurity that are quite unlike the principal substance in molecular composition and structure can be very serious contaminants for certain properties. In other cases, even what are normally insignificant amounts of impurity can seriously affect certain solid-state properties, particularly electrical or magnetic ones, as well as, in some cases, the property of thermal conductivity.[8] In every case, the investigator should record for publication all relevant and necessary information regarding purity.

Measurements of temperature are usually of two kinds: (a) the absolute temperature at which a given observation is being made, and (b) the change in temperature arising from a given input or output of energy or appropriate other phenomenon. For some measurements, as in determining the enthalpy of a given reaction, it is not really necessary to know the change in temperature in degrees C or kelvins, if one is simply comparing unknown energy from the reaction with known energy from a standard reference substance or from measured electrical energy. The change in temperature need be such as to be reproducible with the required precision. The thermometric system used in the investigation should be able to provide absolute temperatures with the required accuracy and precision, based on the International Practical Temperature Scale of 1968 (IPTS-68). In Chapters 3 and 4 we have discussed IPTS-68 and the defining point, the primary fixed points, and the secondary fixed points that can be used to reproduce it. The given thermometric system should be monitored at appropriate times with the proper fixed points to determine its continued suitability for the given investigation. The investigator has responsibility for recording for publication all relevant and necessary information regarding the accuracy of his measurements of temperature.

Absolute determinations of pressure are among the most difficult measurements to make. For this reason, standard reference substances should be used in so far as possible. The scale of pressure has been discussed in detail in Chapter 6. The investigator has responsibility for recording for publication all relevant and necessary information regarding the accuracy of his measurements of pressure.

In certain cases involving a need to know pressures, as in determining the boiling points of a pure substance at different pressures, recourse can be had to the use of a reference substance of which the pressure-temperature relations are known with adequate accuracy. In this way, water can be used as a reference material in determining the boiling points at different pressures of pure chemical compounds.[7,9,10]

In measuring intervals of time, as for determining energy as the product of time and electrical power, or for measurement of viscosity, the investigator can today easily utilize the standard time signals discussed in Chapter 2. In every case, the investigator should record for publication all relevant and necessary information regarding his measurements of time.

In measuring electrical power with high accuracy, the investigator usually measures: (a) the potential drop across the resistance coil through which energy is introduced into the system, and (b) the current through that resistance coil by measuring the potential drop across a standard resistor outside in series with the resistance coil in the system. Such measurements require that the standard cells and standard resistors used have been calibrated against known standards certified by the national standardizing laboratory, as the National Bureau of Standards in the United States. When making such measurements, the investigator has the responsibility of recording for publication all relevant and necessary information regarding his measurements of power.

In certain cases, as in the calorimetric determination of the enthalpy of chemical reactions, it is possible to use standard substances certified by a national standardizing laboratory for use under specified conditions to transfer the unit of energy from the national standardizing laboratory to the laboratory of the investigator. Examples of this procedure are the following: (a) the use of benzoic acid to determine the energy equivalent of a bomb combustion calorimeter;[7,11,12] (b) the use of the reaction of burning hydrogen in oxygen to form water to determine the energy equivalent of an atmospheric pressure flame reaction calorimeter;[7,13] (c) the use of tris(hydroxymethyl) aminomethane for determining the energy equivalent of calorimeters for heats of solution or reaction at atmospheric pressure.[14,15] The use of standard substances in this way obviates the need for setting up costly apparatus for measuring electrical energy with high accuracy.

In measuring lengths, the investigator needs to follow the basic guide lines and standards for such measurements as provided by the national standardizing laboratories, as discussed in Chapter 2. In each case where length is a property to be measured, the investigator should record for publication all relevant and necessary information regarding his apparatus and procedures for measuring length.

For measurements of volume of a given containing vessel, recourse is usually had to the measurement, at a given temperature, of the mass of a given pure substance, of accurately known density, required to fill the vessel. Since such a measurement becomes simply the measurement of mass, the investigator is required to use weights that have been calibrated against weights bearing certified values from a national standardizing laboratory. In each case, the investigator has responsibility for recording all relevant and necessary information pertaining to his measurements of mass. The unit of mass and standards of mass are discussed in Chapter 2.

Not all of the properties mentioned above, as temperature, pressure, time, length, mass, power, and volume, will normally be involved in a significant manner in any one investigation. The investigator should plan his investigation in such a way as to take full advantage of the standard reference substances for various properties that are made available by the national standardizing laboratories, and full advantage of the "substitution" method wherever available. For example, the National Bureau of Standards has available a large number of standard reference substances for calibrating measuring apparatus in absolute units. Mention has already been made of benzoic acid for bomb combustion calorimeters and tris(hydroxymethyl)aminomethane for calorimeters for heats of solution and reaction at constant pressure. Other such standards include the following:[16] substances for measurements of refractive index;[17] substances for measurements of density;[17,18] substances for measurements of temperature (as described in Chapters 4 and 5); etc. In all such cases, the investigator should clearly record all relevant and necessary information pertaining to the reference substances used, the values assigned to them, the procedures for their use, etc.

It is difficult to measure the properties of a substance in terms of absolute units of measurement (discussed in Chapter 2). Whenever possible, the investigator should take full advantage of the calibration services and the standard reference substances available from the national standardizing laboratories in the various countries. In the United States the National Bureau of Standards provides lists of the calibration services and standard reference substances available.[16] By utilizing such calibration services and standard reference substances, the investigator can attain from his measuring system the highest precision of which it is capable, with the accuracy of his work then dependent upon the accuracy of the standards utilized.

In planning his investigation, the experimenter should analyze the work to identify those quantities to be measured which are (a) critical to the investigation, (b) important in a secondary manner, and (c) important only in a peripheral manner. As an example, in determining the heat of combustion of a liquid hydrocarbon utilizing benzoic acid as the standard reference substance to determine the energy equivalent of the bomb combustion calorimeter, the critical items are (1) the purity of the liquid hydrocarbon and the oxygen used for the combustion, (2) the "purity of the reaction," (3) the mass of benzoic acid used in each experiment, and (4) the mass of the liquid hydrocarbon used in each experiment. Since the calorimeter is used in a "substitution" manner, it becomes simply the device to absorb and compare the energy from the combustion of the benzoic acid with the energy from the combustion of the hydrocarbon. The change in temperature in each experiment need not be measured in actual degrees on the temperature scale. The thermometric system needs only to reproduce the same change in temperature with the necessary precision. The actual temperature needs to be known only to the extent compatible with the uncertainty of the final result and the temperature coefficient (ΔC_p) of the heat of the given reaction. The investigator does not need to know the heat capacity of the calorimeter or any of its parts, or of the calorimeter water, etc. He needs only to use the same standard calorimeter system in each experiment, making correction for small differences from experiment to experiment.

In general, we can say that the responsibility of the investigator is to utilize the apparatus available to him, in combination with available standard reference substances and calibrations of his instruments referred through to the standards of the national standardizing laboratory, to produce the most reliable results in his power.

Finally, the investigator must identify exactly the things he measured and describe clearly in quantitative terms the values he has used for the physical constants and atomic masses that are required in the reduction of his observations to produce the final values reported in his publication.

C. RESPONSIBILITY OF THE REVIEWER AND APPRAISER

The responsibility of the reviewer and appraiser is to extract from the published record of the given investigation the maximum amount of numerical information with a reliable indication of the uncertainty associated with it, preserving all the accuracy and precision resident in the record.

This means that the reviewer and appraiser must check the record as to how each of the possible variables of measurement was calibrated. He must obtain information, if available, on the calibration of the thermometric system, the calibration of the pressure system, the calibration of the system for measuring mass, the calibration of the system for measuring time, etc. If reference substances are used in measuring properties such as density, volume (by mass), refractive index, heat of combustion, etc., all relevant information must be extracted.

The reviewer and appraiser must identify precisely those quantities that were measured by the investigator and ascertain the values of such physical constants and atomic masses that may have been used by the investigator in conjunction with his own measured quantities to obtain the final values reported in the publication.

In accordance with such changes as may have occurred in the internationally accepted values of the physical constants (Chapter 12) and atomic masses (Chapter 8), and any known changes that may have occurred in the internationally accepted values for any other quantities that may have been assumed by the investigator, as in measurements of temperature on IPTS-48 (Chapter 5) or in the values of substances for measurements of pressure (Chapter 6), the reviewer and appraiser must convert all the observations, constants, etc., to the present-day basis.

For example, in connection with measurements of energy performed before 1910, between 1910 and 1930, between 1930 and 1948, or subsequent to 1948, the reviewer and appraiser carefully identify what the investigator actually measured, as energy in terms of the heat capacity of water at a given temperature, or in terms of 1908 international joules, or in terms of absolute joules (see the discussion in Chapters 9 and 10).

D. DISCUSSION

In this chapter an effort has been made to outline in broad form, with occasionally some detail, the responsibility of the investigator in recording for publication all relevant and necessary information pertaining to the measurements and calculations of his investigation, and the responsibility of the reviewer and appraiser in extracting from the record all the relevant information of the work, preserving all the accuracy and precision resident in the investigation, and converting the results of that investigation into the units and constants currently internationally accepted.

If each investigator and each reviewer and appraiser does his job properly, the scientific and technical community of the world will be well served.

At this point it may be helpful to note the report of the CODATA Task Group on Publication of Data in the Primary Literature[3] includes the following topics: description of experimental procedures, reduction of experimental data, and presentation of numerical results. Also included[3] is a list of available guides for reporting data in various disciplines, including chemical kinetics, crystallography, spectral data of all sorts, thermal conductivity, thermochemistry and thermodynamics, chromatography, and thermal analysis.

E. REFERENCES

1. Rossini, F. D., *Report on the Presentation and Analysis of Scientific Data in the Literature,* Office of Critical Tables, National Research Council, Washington, D.C., 1958.
2. A guide to Procedures for the publication of thermodynamic data, *Pure Appl. Chem.,* 29, 397, 1972. Report of a subcommittee (Edgar F. Westrum, Jr., Chairman) of the Commission on Thermodynamics and Thermochemistry of the International Union of Pure and Applied Chemistry.
3. Guide for the Presentation in the Primary Literature of Numerical Data Derived from Experiments, Report of the CODATA Task Group on Publication of Data in the Primary Literature (Edgar F. Westrum, Jr., Chairman). *UNESCO-UNISIST Guide,* UNESCO, Division of Scientific and Technological Information and Documentation, 7 Place de Fontenoy, 75700 Paris, France. Also in *NSRDS News,* Office of Standard Reference Data, National Bureau of Standards, Washington, D.C. 20234 (February 1974).
4. **Rossini, F. D., Pitzer, K. S., Arnett, R. L., Braun, R. M., and Pimentel, G. C.,** *Selected Values of Physical and Thermodynamic Properties of Hydrocarbons and Related Compounds,* Carnegie Press, Carnegie Institute of Technology (now Carnegie-Mellon University), Pittsburgh, 1953.
5. **Mair, B. J., Glasgow, A. R., Jr., and Rossini, F. D.,** *J. Res. Nat. Bur. Stand.,* 26, 591, 1941.
6. **Taylor, W. J. and Rossini, F. D.,** *J. Res. Nat. Bur. Stand.,* 32, 197, 1944.
7. **Rossini, F. D.,** *Chemical Thermodynamics,* John Wiley and Sons, New York, 1950.
8. **Touloukian, Y.,** Thermophysical Properties Research Center, Purdue University, Lafayette, Ind.
9. **Willingham, C. B., Taylor, W. J., Pignocco, J. M., and Rossini, F. D.,** *J. Res. Nat. Bur. Stand.,* 35, 219, 1945.
10. **Forziati, A. F., Norris, W. R., and Rossini, F. D.,** *J. Res. Nat. Bur. Stand.,* 43, 555, 1949.
11. **Jessup, R. S.,** *J. Res. Nat. Bur. Stand.,* 29, 247, 1942.
12. **Prosen, E. J. and Rossini, F. D.,** *J. Res. Nat. Bur. Stand.,* 33, 439, 1944.
13. **Rossini, F. D.,** *J. Res. Nat. Bur. Stand.,* 6, 1, 1931; 7, 329, 1931.
14. **Prosen, E. J. and Kilday, M. V.,** *J. Res. Nat. Bur. Stand.,* 77A, 581, 1973.
15. **Melaugh, R. A. and Rossini, F. D.,** Department of Chemistry, Rice University, Houston, Report in process of publication.
16. *Standard Reference Materials,* NBS Special Publication 260, National Bureau of Standards, Washington, D.C. 20234.
17. **Forziati, A. F. and Rossini, F. D.,** *J. Res. Nat. Bur. Stand.,* 43, 473, 1949.
18. **Forziati, A. F., Mair, B. J., and Rossini, F. D.,** *J. Res. Nat. Bur. Stand.,* 35, 513, 1945.

AUTHOR INDEX

M

N

O

P

R

S

T

V

W

Y

Z

INDEX